Ben Stacy Jerrik (Ed.)

Brampton Ash

AF307157

Ben Stacy Jerrik (Ed.)

Brampton Ash

Northamptonshire, Corby, St. Mary the Virgin, Brampton Ash

Part Press

Imprint

Permission is granted to copy, distribute and/or modify this document under the terms of the GNU Free Documentation License, Version 1.2 or any later version published by the Free Software Foundation; with no Invariant Sections, with the Front-Cover Texts, and with the Back- Cover Texts. A copy of the license is included in the section entitled "GNU Free Documentation License".

All parts of this book are extracted from Wikipedia, the free encyclopedia (www.wikipedia.org).

You can get detailed informations about the authors of this collection of articles at the end of this book. The editors (Ed.) of this book are no authors. They have not modified or extended the original texts.

Pictures published in this book can be under different licences than the GNU Free Documentation License. You can get detailed informations about the authors and licences of pictures at the end of this book.

The content of this book was generated collaboratively by volunteers. Please be advised that nothing found here has necessarily been reviewed by people with the expertise required to provide you with complete, accurate or reliable information. Some information in this book maybe misleading or wrong. The Publisher does not guarantee the validity of the information found here. If you need specific advice (f.e. in fields of medical, legal, financial, or risk management questions) please contact a professional who is licensed or knowledgeable in that area.

Any brand names and product names mentioned in this book are subject to trademark, brand or patent protection and are trademarks or registered trademarks of their respective holders. The use of brand names, product names, common names, trade names, product descriptions etc. even without a particular marking in this works is in no way to be construed to mean that such names may be regarded as unrestricted in respect of trademark and brand protection legislation and could thus be used by anyone.

Cover image: www.ingimage.com
Concerning the licence of the cover image please contact ingimage.

Publisher:
Part Press is a trademark of
International Book Market Service Ltd., 17 Rue Meldrum, Beau Bassin, 1713-01 Mauritius
Email: info@bookmarketservice.com
Website: www.bookmarketservice.com

Published in 2012

Printed in: U.S.A., U.K., Germany. This book was not produced in Mauritius.

ISBN: 978-613-6-25953-6

Contents

Articles

LE_postcode_area	1
Northamptonshire	4
Corby	24
St._Mary_the_Virgin,_Brampton_Ash	33
A427_road	36
Desborough	38
Kettering	42
Market_Harborough	51

References

Article Sources and Contributors	60
Image Sources, Licenses and Contributors	61

LE_postcode_area

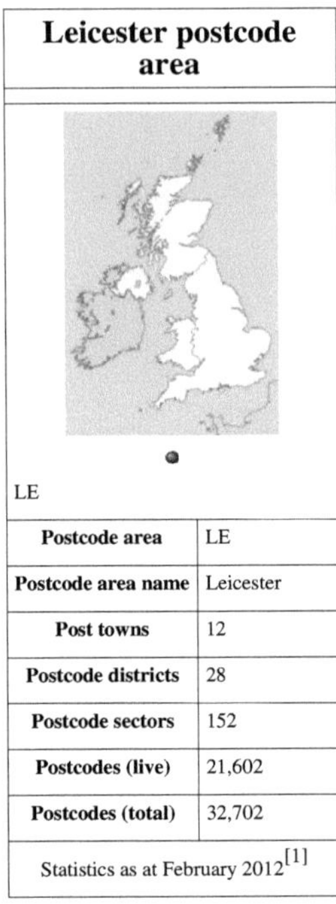

Leicester postcode area	
Postcode area	LE
Postcode area name	Leicester
Post towns	12
Postcode districts	28
Postcode sectors	152
Postcodes (live)	21,602
Postcodes (total)	32,702
Statistics as at February 2012[1]	

The **LE postcode area**, also known as the **Leicester postcode area**[2] , is a group of postcode districts around Coalville, Hinckley, Leicester, Loughborough, Lutterworth, Market Harborough, Oakham, Melton Mowbray & Wigston in Leicestershire, Rutland and Nottinghamshire in central England.

Coverage

The approximate coverage of the postcode districts:

Postcode district	Post town	Coverage	Local authority area
LE1	LEICESTER	Leicester	
LE2	LEICESTER	Oadby, Knighton, Highfields, Aylestone, Glen Parva	
LE3	LEICESTER	Braunstone, Glenfield, New Parks, Groby Road (A50), Leicester Forest East, Westcotes	
LE4	LEICESTER	Birstall, Belgrave, Beaumont Leys, Thurmaston	
LE5	LEICESTER	Hamilton, Thurnby Lodge, Evington	
LE6	LEICESTER	Ratby, Groby, Newtown Linford	
LE7	LEICESTER	Scraptoft, Anstey, Billesdon, Gaddesby, Hungarton, Rearsby, Tilton on the Hill, Tugby, Cropston, Thurcaston, Rothley, Barkby, Syston	
LE8	LEICESTER	Great Glen, Fleckney, Kibworth, Peatling Magna, Countesthorpe, Whetstone	

LE9	LEICESTER	Stoney Stanton, Cosby, Huncote, Croft, Desford, Newbold Verdon, Kirkby Mallory, Earl Shilton, Barwell, Sapcote, Sutton in the Elms, Broughton Astley, Thurlaston	
LE10	HINCKLEY		
LE11	LOUGHBOROUGH	Loughborough, Charnwood	Charnwood
LE12	LOUGHBOROUGH	East Leake, West Leake, Sutton Bonington, Mountsorrel, Shepshed, Belton, Quorn, Sileby, Wymeswold	Charnwood; Rushcliffe
LE13	MELTON MOWBRAY		
LE14	MELTON MOWBRAY	Brooksby, Harby, Hoby, Ragdale, Rotherby, Scalford, Somerby, Stonesby, Waltham on the Wolds, Wymondham	Melton
LE15	OAKHAM	Empingham, Manton, Thistleton, Uppingham, Whissendine, Langham	Rutland
LE16	MARKET HARBOROUGH	East Langton, Hallaton, Market Harborough, Medbourne, Braybrooke	Harborough, Rutland, Kettering
LE17	LUTTERWORTH	Leire, Lutterworth, Swinford, Bitteswell, Ullesthorpe	
LE18	WIGSTON		
LE19	LEICESTER	Narborough, Enderby	
LE21	LEICESTER		
LE41	LEICESTER		
LE55	LEICESTER		
LE65	ASHBY-DE-LA-ZOUCH	Ashby-de-la-Zouch, Boundary, Calke, Smisby, Willesley, Worthington,	
LE67	COALVILLE, IBSTOCK, MARKFIELD		
LE87	LEICESTER		
LE94	LEICESTER		
LE95	LEICESTER		

Adjacent areas

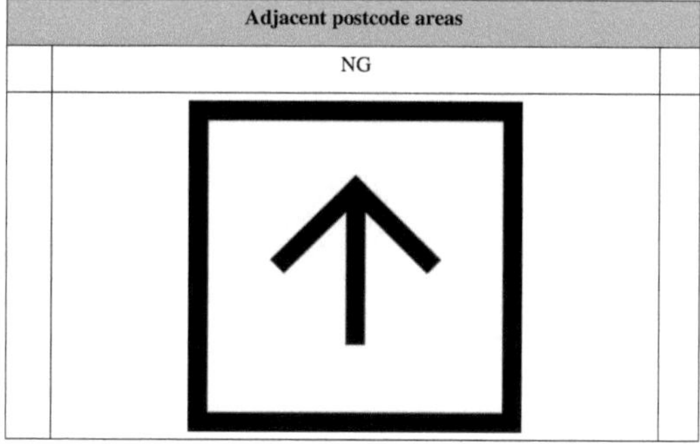

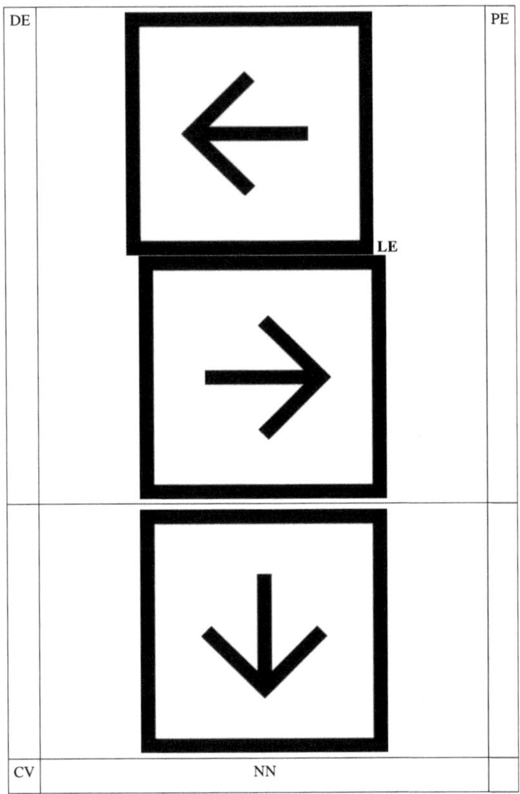

Clockwise from the east, the adjacent postcode areas are: PE (Peterborough), NN (Northampton), CV (Coventry), DE (Derby), NG (Nottingham).

See also

- List of postcode areas in the United Kingdom

References

[1] "ONS Postcode Directory Version Notes" (http://www.ons.gov.uk/ons/guide-method/geography/products/postcode-directories/-nspp-/onspd-user-guide-and-version-notes.zip) (ZIP). *National Statistics Postcode Products* (http://www.ons.gov.uk/ons/guide-method/geography/products/postcode-directories/-nspp-/index.html). Office for National Statistics. February 2012. Table 2. . Retrieved 21 April 2012. Coordinates from mean of unit postcode points, "Code-Point Open" (http://www.ordnancesurvey.co.uk/oswebsite/products/code-point-open/). *OS OpenData* (http://www.ordnancesurvey.co.uk/oswebsite/docs/licences/os-opendata-licence.pdf). Ordnance Survey. February 2012. . Retrieved 21 April 2012.
[2] Royal Mail, *Address Management Guide*, (2004)

Northamptonshire

<table>
<tr><td colspan="2" align="center">Northamptonshire</td></tr>
<tr><td colspan="2" align="center">
Flag</td></tr>
<tr><td colspan="2">Motto of County Council: Rosa concordia signum (The rose: emblem of harmony)</td></tr>
<tr><td colspan="2" align="center">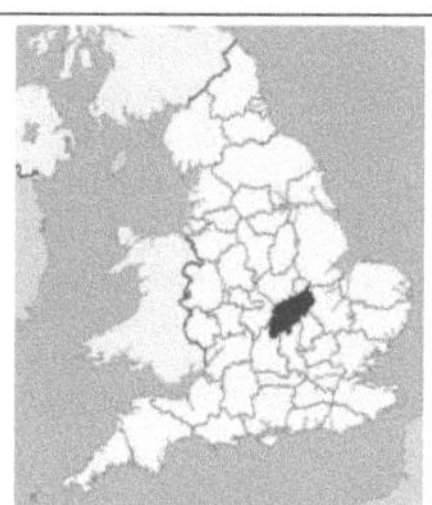</td></tr>
<tr><td colspan="2" align="center">Geography</td></tr>
<tr><td>Status</td><td>Ceremonial & Non-metropolitan county</td></tr>
<tr><td>Region</td><td>East Midlands</td></tr>
<tr><td>Area
- Total
- Admin. council</td><td>Ranked 24th
2364 km^2 (unknown operator: u'strong' sq mi)
Ranked 22nd</td></tr>
<tr><td>Admin HQ</td><td>Northampton</td></tr>
<tr><td>ISO 3166-2</td><td>GB-NTH</td></tr>
<tr><td>ONS code</td><td>34</td></tr>
<tr><td>NUTS 3</td><td>UKF23</td></tr>
<tr><td colspan="2" align="center">Demography</td></tr>
<tr><td>Population
- Total (2010 est.)
- Density
- Admin. council</td><td>Ranked 33rd
687,300
291 /km^2 (unknown operator: u'strong' /sq mi)
Ranked 15th</td></tr>
<tr><td>Ethnicity</td><td>95.1% White
2.0% South Asian
1.2% Black British.</td></tr>
<tr><td colspan="2" align="center">Politics</td></tr>
</table>

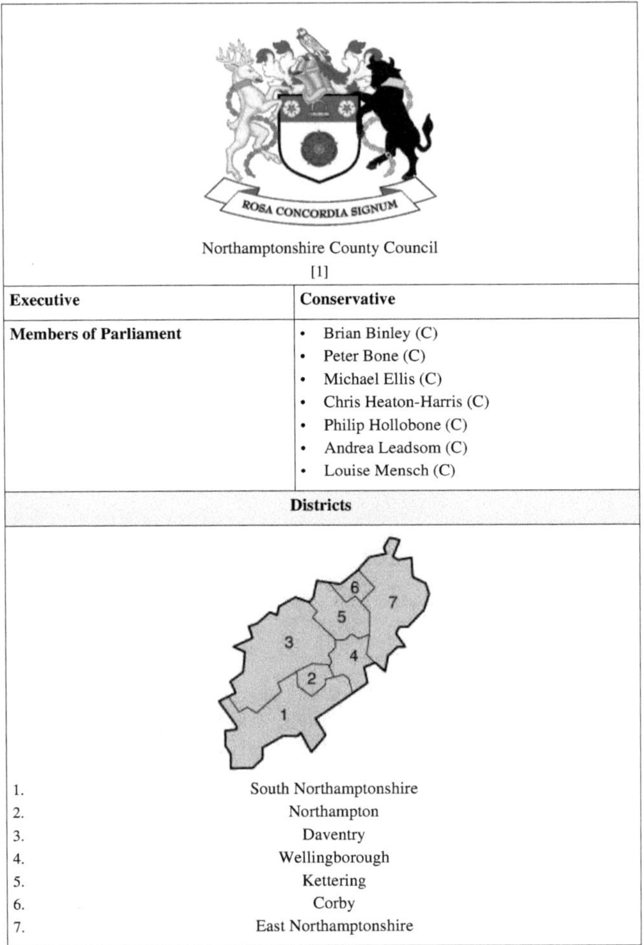

Northamptonshire County Council	
[1]	
Executive	**Conservative**
Members of Parliament	• Brian Binley (C) • Peter Bone (C) • Michael Ellis (C) • Chris Heaton-Harris (C) • Philip Hollobone (C) • Andrea Leadsom (C) • Louise Mensch (C)
Districts	

1.	South Northamptonshire
2.	Northampton
3.	Daventry
4.	Wellingborough
5.	Kettering
6.	Corby
7.	East Northamptonshire

Northamptonshire (◄» /nɔrˈθæmptənʃər/ or /nɔrθˈhæmptənʃɪər/; archaically, the **County of Northampton**; abbreviated **Northants.** or **N/hants**) is a landlocked ceremonial county in the East Midlands region of England. Its population is 629,676 as at the 2001 census. It has boundaries with eight other ceremonial counties: Warwickshire to the west, Leicestershire and Rutland to the north, Cambridgeshire to the east, Bedfordshire to the south-east, Buckinghamshire to the south, Oxfordshire to the south-west and Lincolnshire to the north-east – England's shortest county boundary at 19 metres (**unknown operator: u'strong'** yd).[2] The county seat is Northampton. Other large population centres include Kettering, Corby, Wellingborough, Rushden and Daventry.

Northamptonshire's county flower is the cowslip.

History

Much of Northamptonshire's countryside appears to have remained somewhat intractable with regards to early human occupation, resulting in an apparently sparse population and relatively few finds from the Palaeolithic, Mesolithic and Neolithic periods.[3] In about 500 BC the Iron Age was introduced into the area by a continental people in the form of the Hallstatt culture,[4] and over the next century a series of hill-forts were constructed at Arbury Camp, Rainsborough camp, Borough Hill, Castle Dykes, Guilsborough, Irthlingborough, and most notably of all, Hunsbury Hill. There are two more possible hill-forts at Arbury Hill (Badby) and Thenford.[4]

In the 1st century BC, most of what later became Northamptonshire became part of the territory of the Catuvellauni, a Belgic tribe, the Northamptonshire area forming their most northerly possession.[4] The Catuvellauni were in turn conquered by the Romans in 43 AD.[5]

The Roman road of Watling Street passed through the county, and an important Roman settlement, *Lactodorum*, stood on the site of modern-day Towcester. There were other Roman settlements at Northampton, Kettering and along the Nene Valley near Raunds. A large fort was built at Longthorpe.[4]

After the Romans left, the area eventually became part of the Anglo-Saxon kingdom of Mercia, and Northampton functioned as an administrative centre. The Mercians converted to Christianity in 654 AD with the death of the pagan king Penda.[6] From about 889 the area was conquered by the Danes (as at one point almost all of England was, except for Athelney marsh in Somerset) and became part of the Danelaw - with Watling Street serving as the boundary - until being recaptured by the English under the Wessex king Edward the Elder, son of Alfred the Great, in 917. Northamptonshire was conquered again in 940, this time by the Vikings of York, who devastated the area, only for the county to be retaken by the English in 942.[7] Consequently, it is one of the few counties in England to have both Saxon and Danish town-names and settlements.

The county was first recorded in the Anglo-Saxon Chronicle (1011), as *Hamtunscire*: the *scire* (shire) of *Hamtun* (the homestead). The "North" was added to distinguish Northampton from the other important *Hamtun* further south: Southampton - though the origins of the two names are in fact different.[8]

Rockingham Castle was built for William the Conqueror[9] and was used as a Royal fortress until Elizabethan times. The now-ruined Fotheringhay Castle was used to imprison Mary, Queen of Scots, before her execution.[10] In 1460, during the Wars of the Roses, the Battle of Northampton took place and King Henry VI was captured.[11]

George Washington, the first President of the United States of America, was born into the Washington family who had migrated to America from Northamptonshire in 1656. George Washington's great-great-great-great-great grandfather, Lawrence Washington, was Mayor of Northampton on several occasions and it was he who bought Sulgrave Manor from Henry VIII in 1539. It was George Washington's great-grandfather, John Washington, who emigrated in 1656 from Northants to Virginia. Before Washington's ancestors moved to Sulgrave, they lived in Warton, Lancashire.[12]

John Speed's 17th century map of Northamptonshire

During the English Civil War, Northamptonshire strongly supported the Parliamentarian cause, and the Royalist forces suffered a crushing defeat at the Battle of Naseby in 1645 in the north of the county. King Charles I was imprisoned at Holdenby House in 1647.[13]

In 1823 Northamptonshire was said to "[enjoy] a very pure and wholesome air" because of its dryness and distance from the sea. Its livestock were celebrated: "Horned cattle, and other animals, are fed to extraordinary sizes: and many horses of the large black breed are reared."[14]

Nine years later, the county was described as "a county enjoying the reputation of being one of the healthiest and pleasantest parts of England" although the towns were "of small importance" with the exceptions of Peterborough and Northampton. In summer, the county hosted "a great number of wealthy families... country seats and villas are to be seen at every step."[15] Northamptonshire is still referred to as the county of "spires and squires" because of the numbers of stately homes and ancient churches.[16]

In the 18th and 19th centuries, parts of Northamptonshire and the surrounding area became industrialised. The local specialisation was shoemaking and the leather industry and by the end of the 19th century it was almost definitively the boot and shoe making capital of the world. In the north of the county a large ironstone quarrying industry developed from 1850.[17] During the 1930s, the town of Corby was established as a major centre of the steel industry. Much of Northamptonshire nevertheless remains largely rural.

Corby was designated a new town in 1950[18] and Northampton followed in 1968.[19] As of 2005 the government is encouraging development in the South Midlands area, including Northamptonshire.[20]

Peterborough

The Soke of Peterborough was historically associated with and considered part of Northamptonshire, as the county diocese is focused upon the cathedral there.[21] However, Peterborough had its own county council, and in 1965 was merged with the neighbouring small county of Huntingdonshire.[22] Under the Local Government Act 1972 the city of Peterborough became a district of Cambridgeshire.[23]

Geography

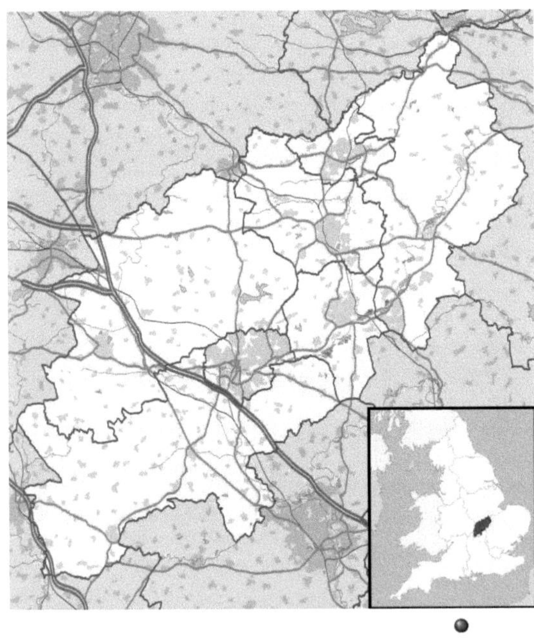

Corby

Daventry

Rushden

Thrapston

Brackley

Oundle

Desborough

Towcester

Irthlingborough

Kings Sutton

Brixworth

Raunds

Silverstone

Banbury

Market Harborough

Milton Keynes

Rugby

Notable places in and around Northamptonshire

Northamptonshire is a landlocked county located in the southern part of the East Midlands region[24] which is sometimes known as the South Midlands. The county contains the watershed between the River Severn and The Wash while several important rivers have their sources in the north-west of the county, including the River Nene, which flows north-eastwards to The Wash, and the "Warwickshire Avon", which flows south-west to the Severn. In 1830 it was boasted that "not a single brook, however insignificant, flows into it from any other district".[25] The highest point in the county is Arbury Hill at 225 metres (**unknown operator: u'strong' ft).[26]

Kilworth Wharf on the Grand Union Canal

There are several towns in the county with Northampton being the largest and most populous. At the time of the 2008 estimates, a population of 685,000 lived in the county with 205,200 living in Northampton. The table below shows all towns with over 9,000 inhabitants.

Rank	Town	Population	Borough/District council
1	Northampton	205,200 (2008)	Northampton Borough Council
2	Kettering	51,063 (2001)[27]	Kettering Borough Council
3	Corby	49,222 (2001)[27]	Corby Borough Council
4	Wellingborough	46,959 (2001)[27]	Borough Council of Wellingborough
5	Rushden	25,849 (2001)	East Northamptonshire District Council
6	Daventry	22,367 (2001)	Daventry District Council
7	Brackley	13,331 (2001)	South Northamptonshire District Council

As of 2010 there are 16 settlements in Northamptonshire with a town charter:

- Brackley, Burton Latimer, Corby, Daventry, Desborough, Higham Ferrers, Irthlingborough, Kettering, Northampton, Oundle, Raunds, Rothwell, Rushden, Towcester, Thrapston and Wellingborough.

Climate

Like the rest of the British Isles, Northamptonshire has an oceanic climate (Köppen climate classification). The table below shows the average weather for Northamptonshire from the Moulton weather station.

Governance

Northamptonshire, like most English counties, is divided into a number of local authorities. The seven borough/district councils cover 15 towns and hundreds of villages. The county has a two-tier structure of local government and an elected county council based in Northampton, and is also divided into seven districts each with their own district or borough councils:[28]

Council	Where based
Corby Borough Council	Corby
Daventry District Council	Daventry
East Northamptonshire District Council	Thrapston
Kettering Borough Council	Kettering
Northampton Borough Council	Northampton
South Northamptonshire District Council	Towcester
Borough Council of Wellingborough	Wellingborough

Northampton itself is the most populous urban district in England not to be administered as a unitary authority (even though several smaller districts are unitary). During the 1990s local government reform, Northampton Borough Council petitioned strongly for unitary status, which led to fractured relations with the County Council.

Before 1974, the Soke of Peterborough was considered geographically part of Northamptonshire, although it had had a separate county council since the late 19th Century and separate Quarter Sessions courts before then. Now part of Cambridgeshire, the city of Peterborough became a unitary authority in 1998, but it continues to form part of that county for ceremonial purposes.[29]

National representation

Northamptonshire returns seven members of Parliament, who all are part of the Conservative Party.[30]

Constituency	Member of Parliament	Political party
Corby	Louise Mensch	Conservative
Daventry	Chris Heaton-Harris	Conservative
Kettering	Philip Hollobone	Conservative
Northampton North	Michael Ellis	Conservative
Northampton South	Brian Binley	Conservative
Northamptonshire South	Andrea Leadsom	Conservative
Wellingborough & Rushden	Peter Bone	Conservative

From 1993 until 2005, Northamptonshire County Council,[31] for which each of the 73 electoral divisions in the county elect a single councillor, had been held by the Labour Party; previously it had been under no overall control since 1981. The councils of the rural districts – Daventry, East Northamptonshire, and South Northamptonshire – are strongly Conservative, whereas the political composition of the urban districts is more mixed. At the 2003 local elections, Labour lost control of Kettering, Northampton, and Wellingborough, retaining only Corby. Elections for the entire County Council are held every four years – the last were held on 5 May 2005 when control of the County Council changed from the Labour Party to the Conservatives. The County Council uses a leader and cabinet executive system and abolished its area committees in April 2006.

Economy

Historically, Northamptonshire's main industry was the manufacture of boots and shoes.[32] Many of the manufacturers closed down in the Thatcher era which in turn left many county people unemployed. Although R Griggs and Co Ltd, the manufacturer of Dr. Martens, still has its UK base in Wollaston near Wellingborough,[33] the shoe industry in the county is now nearly gone. Large employers include the breakfast cereal manufacturers Weetabix, in Burton Latimer, the Carlsberg brewery in Northampton, Avon Products, Siemens, Barclaycard, Saxby Bros Ltd and Golden Wonder.[34] [35] In the west of the county is the Daventry International Railfreight Terminal;[36] which is a major rail freight terminal located on the West

Silverstone adds millions every year to the local economy - Kimi Räikkönen testing for McLaren at Silverstone in April 2006

Coast Main Line near Rugby. Wellingborough also has a smaller railfreight depot[37] on Finedon Road, called Nelisons sidings.[38]

This is a chart of trend of the regional gross value added of Northamptonshire at current basic prices in millions of British Pounds Sterling (correct on 21 December 2005):[39]

Year	Regional Gross Value Added[40]	Agriculture[41]	Industry[42]	Services[43]
1995	6,139	112	2,157	3,870
2000	9,743	79	3,035	6,630
2003	10,901	90	3,260	7,551

The region of Northamptonshire, Oxfordshire and the South Midlands has been described as "Motorsport Valley... a global hub" for the motor sport industry.[44] [45] The Mercedes GP[46] and Force India[47] Formula One teams have their bases at Brackley and Silverstone respectively, while Cosworth[48] and Mercedes-Benz High Performance Engines[49] are also in the county at Northampton and Brixworth.

International motor racing takes place at Silverstone Circuit[50] and Rockingham Motor Speedway;[51] Santa Pod Raceway is just over the border in Bedfordshire but has a Northants postcode.[52] A study commissioned by Northamptonshire Enterprise Ltd (NEL) reported that Northamptonshire's motorsport sites attract more than 2.1 million visitors per year who spend a total of more than £131 million within the county.[53]

Milton Keynes and South Midlands Growth area

Northamptonshire forms part of the Milton Keynes and South Midlands Growth area which also includes Milton Keynes, Aylesbury Vale and Bedfordshire. This area has been identified as an area which is due to have tens of thousands additional homes built between 2010-2020. In North Northamptonshire (Boroughs of Corby, Kettering, Wellingborough and East Northants), over 52,000 homes are planned or newly-built and 47,000 new jobs are also planned.[54] In West Northamptonshire (boroughs of Northampton, Daventry and South Northants), over 48,000 homes are planned or newly-built and 37,000 new jobs are planned.[55] To overlook the planned developments, two urban regeneration companies have been created: North Northants Development Company (NNDC)[54] and the West Northamptonshire Development Corporation.[55] The NNDC launched a controversial[56] campaign called *North Londonshire* to attract people from London to the county.[57] There is also a county-wide tourism campaign with the slogan *Northamptonshire, Let yourself grow.*[58]

Education

Northamptonshire County Council operates a complete comprehensive system with 42 state secondary schools.[59] The county's music and performing arts service provides peripatetic music teaching to schools. It also supports 15 local Saturday morning music and performing arts centres around the county[60] and provides a range of county-level music groups.[61]

Colleges

There are seven colleges across the county, with the Tresham College of Further and Higher Education having four campuses in three towns: Corby, Kettering and Wellingborough.[62] Tresham provides further education and offers vocational courses, GCSEs and A Levels.[63] It also offers Higher Education options in conjunction with several universities.[64] Other colleges in the county are: Fletton House, Knuston Hall, Moulton College, Northampton College, Northampton New College and The East Northamptonshire College.

University

Northamptonshire has one University, the University of Northampton. It has two campuses 2.5 miles (**unknown operator: u'strong' km**) apart and 10,000 students.[65] It offers courses for needs and interests from foundation and undergraduate level to postgraduate, professional and doctoral qualifications. Subjects include traditional arts, humanities and sciences subjects, as well as entrepreneurship, product design and advertising.[66]

Healthcare

Hospitals

Northampton has several National Health Service branches, the main acute NHS hospitals in the county being Northampton and Kettering General Hospitals. In the south-west of the county, the town of Brackley and surrounding villages are serviced by the Horton General Hospital in Banbury in neighbouring Oxfordshire for acute medical needs. A similar arrangement is in place for the town of Oundle and nearby villages, served by Peterborough District Hospital.

In February 2011 a new satellite out-patient centre opened at Nene Park, Irthlingborough to provide over 40,000 appointments a year, as well as a minor injury unit to serve Eastern Northamptonshire. This was opened to relieve pressure off Kettering General Hospital, and has also replaced the dated Rushden Memorial Clinic which provided at the time about 8,000 appointments a year, when open. [67]

Water contamination

In June 2008, Anglian Water found traces of Cryptosporidium in water supplies of Northamptonshire. The local reservoir at Pitsford was investigated and a European Rabbit which had strayed into it was found,[68] causing the problem. About 250,000 residents were affected;[69] by 14 July 2008, 13 cases of cryptosporidiosis attributed to water in Northampton had been reported.[70] Following the end of the investigation, Anglian Water lifted its boil notice for all affected areas on 4 July 2008.[71] Anglian Water revealed that it will pay up to £30 per household as compensation for customers hit by the water crisis.[72]

Transport

The gap in the hills at Watford Gap meant that many south-east to north-west routes passed through Northamptonshire. The Roman Road Watling Street (now part of the A5) passes through here, as did later canals, railways and major roads.

Brackley bypass on the A43

Roads

Major national roads including the M1 motorway (London to Leeds) and the A14 (Rugby to Ipswich), provide Northamptonshire with transport links, both north–south and east–west. The A43 joins the M1 to the M40 motorway, passing through the south of the county to the junction west of Brackley, and the A45 links Northampton with Wellingborough and Peterborough.

The county road network, managed by Northamptonshire County Council includes the A45 west of the M1 motorway, the A43 between Northampton and the county boundary near Stamford, the A361 between Kilsby and Banbury (Oxon) and all B, C and Unclassified Roads. Since 2009 these highways have been managed on behalf of the county council by MGWSP, a joint venture between May Gurney and WSP.

Rivers and canals

Further information: Category:Rivers of Northamptonshire

Two major canals – the Oxford and the Grand Union – join in the county at Braunston. Notable features include a flight of 17 locks on the Grand Union at Rothersthorpe, the canal museum at Stoke Bruerne, and a tunnel at Blisworth which, at 2813 metres (**unknown operator: u'strong'** yd), is the third-longest navigable canal tunnel on the UK canal network.

The Grand Union Canal at Braunston

A branch of the Grand Union Canal connects to the River Nene in Northampton and has been upgraded to a "wide canal" in places and is known as the *Nene Navigation*. It is famous for its guillotine locks.

Railways

Two trunk railway routes, the Midland Main Line and the West Coast Main Line, cross the county. At its peak, Northamptonshire had 75 railway stations. It now has only six, at Northampton and Long Buckby on the West Coast Main Line, Kettering, Wellingborough and Corby on the Midland Main Line, along with King's Sutton, which is a few metres from the boundary with Oxfordshire on the Chiltern Main Line.

A East Midlands Trains service approaching Wellingborough on the Midland Main Line

Before nationalisation of the railways in 1948 and the creation of British Railways, three of the "Big Four" railway companies operated in Northamptonshire: the London, Midland and Scottish Railway, London and North Eastern Railway and Great Western Railway. Only the Southern Railway was not represented. As of 2010 it is served by Virgin Trains, London Midland, Chiltern Railways and East Midlands Trains.

Corby rail history

Corby was described as the largest town in Britain without a railway station.[73] The railway running through the town from Kettering to Oakham in Rutland was previously used only by freight traffic and occasional diverted passenger trains that did not stop at the station. The line through Corby was once part of a main line to Nottingham through Melton Mowbray, but the stretch between Melton and Nottingham was closed in 1968. In the 1980s, an experimental passenger shuttle service ran between Corby and Kettering but was withdrawn a few years later.[74] On 23 February 2009, a new railway station opened, providing direct hourly access to London St Pancras. Following the opening of Corby Station, Rushden then became the largest town in the UK without a direct railway station.

Closed lines and stations

Railway services in Northamptonshire were reduced by the Beeching Axe in the 1960s.[75] Closure of the line connecting Northampton to Peterborough by way of Wellingborough, Thrapston, and Oundle left eastern Northamptonshire devoid of railways. Part of this route was reopened in 1977 as the Nene Valley Railway. A section of one of the closed lines, the Northampton to Market Harborough line, is now the Northampton & Lamport heritage railway, while the route as a whole forms a part of the National Cycle Network, as the Brampton Valley Way.

As early as 1897 Northamptonshire would have had its own Channel Tunnel rail link with the creation of the Great Central Railway, which was intended to connect to a tunnel under the English Channel. Although the complete project never came to fruition, the rail link through Northamptonshire was constructed, and had stations at Charwelton, Woodford Halse, Helmdon and Brackley. It became part of the London and North Eastern Railway in 1923 (and of British Railways in 1948) before its closure in 1966.

Future

In June 2009 the Association of Train Operating Companies (ATOC) recommended opening a new station on the former Irchester railway station site for Rushden, Higham Ferrers and Irchester, called Rushden Parkway.[76] Network Rail is looking at electrifying the Midland Main Line north of Bedford.[77] A open access company has approached Network Rail for services to Oakham in Rutland to London via the county.[77]

The Rushden, Higham and Wellingborough Railway would like to see the railway fully reopen between Wellingborough and Higham Ferrers. As part of the government-proposed High Speed 2 railway line (between London and Birmingham), the High speed railway line will go through the southern part of the county but with no station built.

Buses

Most buses are operated by Stagecoach in Northants and First Northampton. Some town area routes have been named the Corby Star, Connect Kettering, Connect Wellingborough and Daventry Dart; the last three of these routes have route designations that include a letter, such as A, D1, W1, W2, and so on.[78] [79]

Airports

Sywell Aerodrome, on the edge of Sywell village, has three grass runways and one concrete all weather runway. It is however only 1000 metres and therefore cannot be served by passenger jets as of yet.[80]

Sywell Aerodrome

Media

Newspapers

The two main newspapers in the county are the Northamptonshire Evening Telegraph and the Northampton Chronicle & Echo.

Television

BBC regions

Most of Northamptonshire is served by the BBC's East region which is based in Norwich. The regional news television programme, **BBC Look East**, provides local news across the East of England, Milton Keynes and most of Northamptonshire. An opt-out in *Look East* covers

BBC Radio Northampton's Broadcasting House

the west part of the region only, broadcast from Cambridge. This area also is covered by the BBC's **The Politics Show: East** and **Inside Out: East**. A small part of the northern part of the county is covered by BBC East Midlands's regional news **BBC East Midlands Today**, while a small part of South Northamptonshire is covered by BBC Oxford's regional news **BBC Oxford News** which is part of the BBC South Today programme.

ITV regions

Most of Northamptonshire is covered by ITV's Anglia region (which broadcasts **Anglia Today/Tonight**); in the south-west of the county, primarily Brackley and the surrounding villages, broadcasts can be received from the Oxford transmitter which broadcasts ITV Meridian's **Meridian Today/Tonight**.

Radio

BBC Radio Northampton, broadcasts on two FM frequencies: 104.2 MHz for the south and west of the county (including Northampton and surrounding area) and 103.6 MHz for the north of the county (including Kettering, Wellingborough and Corby). BBC Radio Northampton is located in Abington Street, Northampton. These services are broadcast from the Sandy Heath transmitter in Bedfordshire.

There are three commercial radio stations in the county. The former *Kettering and Corby Broadcasting Company (KCBC)* station is now called Connect Radio (97.2 and 107.4 MHZ FM), following a merger with the Wellingborough-based station of the same name. While both Heart Northants (96.6 MHz FM) and AM station Gold (1557 kHz) air very little local content as they form part of a national network. National digital radio is also available in Northamptonshire, though coverage is limited.

Sport

Rugby Union

Northamptonshire has many rugby union clubs. Its premier team, Northampton Saints, competes in the Aviva Premiership and won the European championship in 2000 by defeating Munster for the Heineken Cup, 9-8. Saints are based at the 13,600 capacity Franklin's Gardens ground.

Football

Northamptonshire has several football teams, the most prominent being the League Two side Northampton Town. Other football teams include Kettering Town who are in the Conference National & Corby Town, who are in the Conference North. Wellingborough Town claims to be the sixth oldest club in the country.

Statue inscribed 'They tackled the job' outside Franklin's Gardens

Cricket

Northamptonshire County Cricket Club is in Division Two of the County Championship. Northamptonshire Cricket Club has recently signed overseas professionals such as Sourav Ganguly.

Motor Sport

Silverstone is a major motor racing circuit, most notably used for the British Grand Prix. There is also a dedicated radio station for the circuit which broadcasts on 87.7 FM or 1602 MW when events are taking place. Rockingham Speedway Corby is the largest stadium in the UK with 130,000 seats. It is a US-style elliptical racing circuit (the largest of its kind outside of the US), and is used extensively for all kinds of motor racing events. The Santa Pod drag racing circuit, venue for the FIA European Drag Racing Championships is just across the border in Bedfordshire but has a NN postcode.

Swimming

There are five main swimming clubs in the county: Wellingborough, Northampton, Kettering, Daventry and Rushden. They participate in many competitions. There is also an Olympic sized swimming pool at Corby opened in 2010.

Culture

Rock and pop bands originating in the area have included Bauhaus, The Departure, New Cassettes, Raging Speedhorn and Defenestration.

Places of interest

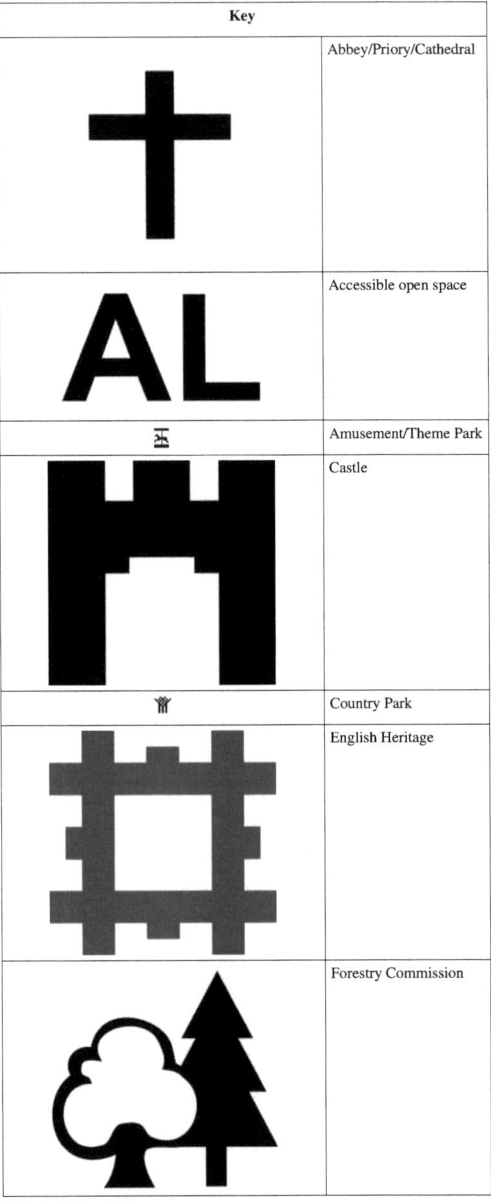

Key	
	Abbey/Priory/Cathedral
	Accessible open space
	Amusement/Theme Park
	Castle
	Country Park
	English Heritage
	Forestry Commission

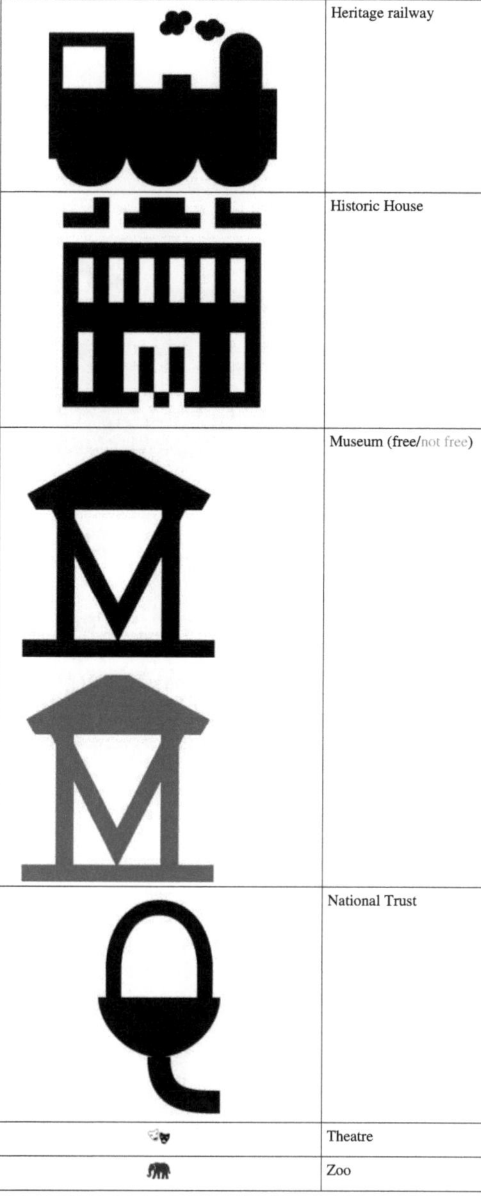

	Heritage railway
	Historic House
	Museum (free/not free)
	National Trust
	Theatre
	Zoo

- 78 Derngate 🏛
- Althorp 🏰
- Barnwell Country Park 🌳
- Barnwell Manor 🏰
- Billing Aquadrome
- Borough Hill Daventry (Iron Age hill fort) **AL**
- Boughton House (home of the Dukes of Buccleuch) 🏰
- Blisworth tunnel
- Brackley

- Knuston Hall 🏰
- Lamport Hall 🏰
- lilford Hall 🏰
- Lyveden New Bield 🔱
- Pitsford Reservoir
- Prebendal Manor House, Nassington 🏰
- Naseby Field
- Northampton Cathedral
- Northampton & Lamport Railway

- Brampton Valley Way (linear park on a disused railway line) **AL**

- Northamptonshire Ironstone Railway

- Brixworth Country Park 🌳
- Burghley House (in the Soke of Peterborough, so formerly in Northants), 🏰
- Canons Ashby House 🔱
- Castle Ashby (home of the Marquess of Northampton), 🏰
- Coton Manor Garden
- Cottesbrooke Hall 🏰
- Daventry Country Park 🌳

- Roadmender - live music venue [81]
- Piddington Roman Villa

- Rockingham Castle 🏰
- Rockingham Forest 🌳
- Rockingham Motor Speedway
- Rushden Hall
- Rushden, Higham and Wellingborough Railway

- Deene Park 🏰

- Rushden Station Railway Museum

- Delapré Abbey

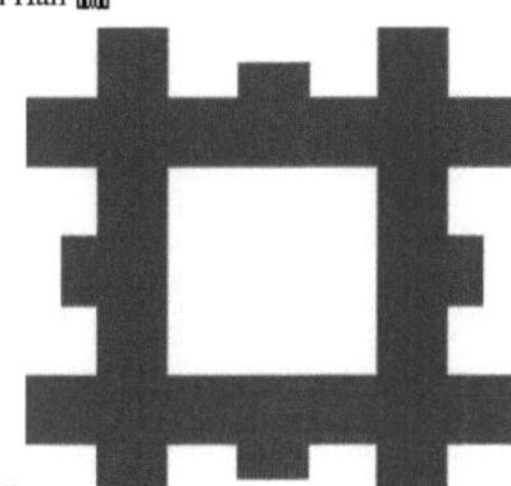

- Derngate and Royal Theatre
- Easton Neston 🏛
- Elton Hall 🏛
- Fermyn Woods Country Park 🌳
- Fotheringhay Castle & Church
- Franklin's Gardens
- Geddington's Eleanor cross
- Holdenby House 🏛
- Irchester Country Park 🌳
- Jurassic Way (long-distance footpath)
- Kelmarsh Hall 🏛
- Kirby Hall

- Rushton Triangular Lodge
- Salcey Forest 🌳
- Silverstone Circuit
- Southwick Hall 🏛
- Stanwick Lakes 🌳
- Stoke Bruerne Canal Museum 🏛
- Sulgrave Manor 🏛
- Summer Leys nature reserve
- Syresham
- Sywell Country Park 🌳
- The Castle Theatre
- Watford Locks
- Wellingborough Museum 🏛

- Whittlewood Forest 🌳
- Wicksteed Park

Annual events

- Gretton Barn dance
- British Grand Prix at Silverstone
- Burghley Horse Trials
- Crick Boat Show
- Hollowell Steam Rally
- Northampton Balloon Festival
- Rothwell Fair
- Rushden Cavalcade
- St Crispin Street Fair
- Wellingborough Carnival
- World Conker Championships

See also

- List of Lord Lieutenants of Northamptonshire
- List of High Sheriffs of Northamptonshire
- Custos Rotulorum of Northamptonshire - List of Keepers of the Rolls
- Northamptonshire (UK Parliament constituency) - Historica list of MPs for the Northamptonshire constituency
- List of places in Northamptonshire
- History of Northamptonshire
- East Midlands
- South Midlands
- Category:People from Northamptonshire

Notes

[1] http://www.northamptonshire.gov.uk/

[2] "Lincolnshire County Council" (http://www.thebythams.org.uk/localgovernment/lincolnshire-cc/index.html). Thebythams.org.uk. 2005-10-24. . Retrieved 2010-09-25.

[3] Greenall (1979) p.19

[4] Greenall (1979) p.20

[5] BBC - History - Tribes of Britain (http://www.bbc.co.uk/history/ancient/british_prehistory/iron_02.shtml). Retrieved 16 August 2009.

[6] Greenall (1979) p.29

[7] Wood, Michael (1986) *The Domesday Quest* p. 90, BBC Books, 1986 ISBN 0-563-52274-7.

[8] Mills, A.D. (1998). A Dictionary of English Place-names. Second Edition. Oxford University Press, Oxford. p256. ISBN 0-19-280074-4

[9] Rockingham Castle (http://www.rockinghamcastle.com/). Retrieved 16 August 2009.

[10] Mott, Allan. BBC - Cambridgeshire - History: Mary Queen of Scots' last days (http://www.bbc.co.uk/cambridgeshire/content/articles/2008/04/28/mary_queen_video_feature.shtml). Retrieved 16 August 2009.

[11] Stearns, Peter N., Langer. William L. The Encyclopedia of world history: ancient, medieval, and modern (http://books.google.co.uk/books?id=MziRd4ddZz4C&pg=PA241&lpg=PA241&dq=Henry+VI+captured+Northampton&source=bl&ots=Y59GhdBeqO&sig=FVzhUUWK-UGWS-wwFgk1-BwMEI8&hl=en&ei=w3yISovTKJbUjAeS4qWiCw&sa=X&oi=book_result&ct=result&resnum=10#v=onepage&q=Henry VI captured Northampton&f=false). Retrieved 16 August 2009.

[12] The Writings of George Washington: Life of Washington (http://books.google.co.uk/books?id=PrfcVGtd0T4C&pg=PA545&lpg=PA545&dq=laurence+washington+warton&source=bl&ots=iyZ1YkQYH8&sig=I4ZUN9rFZ9eZwcNVgWY9rcwvNNY&hl=en&ei=cHqISprWEY3SjAf21YyiCw&sa=X&oi=book_result&ct=result&resnum=4#v=onepage&q=&f=false). Retrieved 16 August 2009.

[13] Edmonds. 1848. Notes on English history for the use of juvenile pupils (http://books.google.com/books?id=nCQEAAAAQAAJ&pg=PA49&dq=Charles+I+imprisoned+Holdenby&as_brr=3#v=onepage&q=&f=false). Retrieved 16 August 2009.

[14] Brookes, R., Whittaker, W.B. *The General Gazetteer, or, Compendious geographical dictionary, in miniature* (http://books.google.com/books?id=XjANAAAAYAAJ&pg=RA2-PA241&dq=geography+of+northamptonshire&lr=&as_brr=1#v=onepage&q=&f=false). 1823. Retrieved 5 September 2009.

[15] Malte-Brun, C. Universal geography: or, A description of all parts of the world (http://books.google.com/books?id=-0gBAAAAYAAJ&pg=PA771&dq=geography+of+northamptonshire&as_brr=1#v=onepage&q=&f=false). 1832. Retrieved 5 September 2009.

[16] Andrews, R., Teller, M. The Rough Guide to Britain (http://books.google.co.uk/books?id=AOt1Hb8MOQUC&pg=PA553&lpg=PA553&dq=Northamptonshire+spires+and+squires&source=bl&ots=BjV7glaFoX&sig=JAOqwiJC80mHBAFb4qNFbQ8Jwqo&hl=en&ei=n56PSqO9DY-SjAePzoz4DQ&sa=X&oi=book_result&ct=result&resnum=8#v=onepage&q=&f=false) 2004. Rough Guides. Retrieved 5 September 2009.

[17] GENUKI: Northamptonshire Genealogy: Bartholomew's Gazetteer of the British Isles, 1887 (http://www.kellner.eclipse.co.uk/genuki/NTH/). 11 August 2008. Retrieved 5 September 2009.

[18] Corby - English Partnerships (http://www.englishpartnerships.co.uk/corby.htm). Retrieved 16 August 2009.

[19] Northampton - English Partnerships (http://www.englishpartnerships.co.uk/northampton.htm). Retrieved 16 August 2009.

[20] Northamptonshire Chamber :: Milton Keynes & South Midlands Growth Plan (http://www.northants-chamber.co.uk/representation/mksm/). Retrieved 16 August 2009.

[21] Peterborough Diocesan Registry (http://www.peterboroughdiocesanregistry.co.uk/). Retrieved 15 August 2009.

[22] The Huntingdon and Peterborough Order 1964 (SI 1964/367), see Local Government Commission for England (1958-1967), *Report and Proposals for the East Midlands General Review Area (Report No.3)*, 31 July 1961 and *Report and Proposals for the Lincolnshire and East Anglia General Review Area (Report No.9)*, 7 May 1965

[23] The English Non-metropolitan Districts (Definition) Order 1972 (SI 1972/2039) Part 5: County of Cambridgeshire

[24] Northamptonshire - Let yourself grow: Media information about Northamptonshire (http://www.explorenorthamptonshire.co.uk/exec/112345/7111/). Retrieved 15 August 2009.

[25] UK Genealogy Archives: Transcript from Pigot & Co's Commercial Directory, 1830 (http://uk-genealogy.org.uk/england/ Northamptonshire/pigot.html). Retrieved 15 August 2009.

[26] Northamptonshire Genealogy: Bartholomew's Gazetteer of the British Isles, 1887 (http://www.kellner.eclipse.co.uk/genuki/NTH/). Retrieved 15 August 2009.

[27] http://www.statistics.gov.uk/statbase/Expodata/Spreadsheets/D8271.csv

[28] Northamptonshire County Council: District and Borough Councils (http://www.northamptonshire.gov.uk/en/Pages/districts.aspx). 2008. Retrieved 22 August 2009.

[29] The Cambridgeshire (City of Peterborough) (Structural, Boundary and Electoral Changes) Order 1996 (http://www.legislation.gov.uk/ uksi/1996/1878/contents/made) (SI 1996/1878), see Local Government Commission for England (1992), *Final Recommendations for the Future Local Government of Cambridgeshire*, October 1994 and *Final Recommendations on the Future Local Government of Basildon & Thurrock, Blackburn & Blackpool, Broxtowe, Gedling & Rushcliffe, Dartford & Gravesham, Gillingham & Rochester upon Medway, Exeter, Gloucester, Halton & Warrington, Huntingdonshire & Peterborough, Northampton, Norwich, Spelthorne and the Wrekin*, December 1995

[30] Northamptonshire County Council: Members of Parliament (http://www.northamptonshire.gov.uk/en/councilservices/council/ mp_mep/pages/mps.aspx). 27 April 2009. Retrieved 22 August 2009.

[31] "Northamptonshire County Council website" (http://www.northamptonshire.gov.uk/). . Retrieved 4 June 2009.

[32] GENUKI: Northamptonshire Genealogy: Bartholomew's Gazetteer of the British Isles (http://www.kellner.eclipse.co.uk/genuki/NTH/). 1887. Retrieved 22 August 2009.

[33] Kellysearch.co.uk: R Griggs & Co. Ltd (http://www.kellysearch.co.uk/gb-company-370000597.html). Retrieved 22 August 2009.

[34] Northamptonshire Chamber: Major Northamptonshire employers (http://www.northants-chamber.co.uk/info/topemployers/#). Retrieved 22 August 2009.

[35] (http://www.wellingborough.gov.uk/downloads/Why_Wellingborough_Final_version.pdf). Retrieved 23 August 2009.

[36] Prologis RFI Dirft Daventry (http://www.prologisrfidirft.co.uk/). Retrieved 22 August 2009.

[37] FirstGBRf: FirstGBRf opens unique depot at Wellingborough (http://www.gbrailfreight.com/news.php?newsid=178). 12 June 2007. Retrieved 22 August 2009.

[38] GB Railfreight: Locations, Wellingborough (http://www.gbrailfreight.com/locations.php?lid=211) Retrieved 11 November 2010

[39] Regional Gross Value Added.*Office for National Statistics* (http://www.statistics.gov.uk/downloads/theme_economy/RegionalGVA. pdf). pp 240–253. 21 December 2005. Retrieved 22 August 2009.

[40] Components may not sum to totals due to rounding

[41] includes hunting and forestry

[42] includes energy and construction

[43] includes financial intermediation services indirectly measured

[44] Coe, N.M., Kelly, P.F, Wai-Chung Yeung, H. Economic geography: a contemporary introduction (http://books.google.co.uk/ books?id=1xV5ZvXYtjUC&pg=PA141&lpg=PA141&dq=motorsport+industry+northamptonshire&source=bl&ots=oM13gCrv3r& sig=RFRZCbtKwvSgU3QbDdGNtforJ6s&hl=en&ei=KIOQSqfCINyMjAfj8-HjDQ&sa=X&oi=book_result&ct=result& resnum=6#v=onepage&q=motorsport industry northamptonshire&f=false). Wiley-Blackwell, 2007. pp 141-143. Retrieved 22 August 2009.

[45] Russell Hotten. Motor racing battles to stay out of pits (http://business.timesonline.co.uk/tol/business/industry_sectors/leisure/ article5983082.ece). TimesOnline. 27 March 2009. Retrieved 22 August 2009.

[46] Official site of Mercedes GP Formula One Team: Contact us (http://www.mercedes-gp.com/includes/privacy.htm). Retrieved 4 March 2010.

[47] Force India F1 Team: Contact us (http://www.forceindiaf1.com/index/page_id/48). Retrieved 22 August 2009.

[48] Cosworth: Contact (http://www.cosworth.com/Default.aspx?id=1089541). Retrieved 22 August 2009.

[49] Mercedes-Benz High Performance Engines Ltd: Contact (http://www.mercedes-benz-hpe.com/hpe/index.htm). Retrieved 22 August 2009.

[50] Silverstone Official Website: Contact Numbers (http://www.silverstone.co.uk/php/ci_overview.html). Retrieved 22 August 2009.

[51] Getting to Rockingham (http://www.rockingham.co.uk/about/gettingto.asp). Retrieved 22 August 2009.

[52] Santa Pod Raceway: Contact/find us/postcode (http://www.santapod.co.uk/g_find.php). Retrieved 22 August 2009.

[53] Motorsport to grow 30% in next decade (http://www.northantset.co.uk/12691/Motorsport-to-grow-3037-in.5401811.jp). Northants Evening Telegraph. 25 June 2009. Retrieved 22 August 2009.

[54] MSKM: North Northants (http://www.mksm.org.uk/area/north-northants.asp) Accessed 2 October 2010

[55] MKSM: West Northants (http://www.mksm.org.uk/area/west-northants.asp) Accessed 2 October 2010

[56] Northants Evening Telegraph: Come to North Londonshire (http://www.northantset.co.uk/news/Come-to-North-Londonshire.6370328. jp) Accessed 2 October 2010

[57] North Londonshire: home page (http://www.northlondonshire.co.uk/) Accessed 2 October 2010

[58] Let yourself grow: home page (http://www.letyourselfgrow.com/) Accessed 2 October 2010

[59] Northamptonshire County Council: Northamptonshire Schools Directory (http://www3.northamptonshire.gov.uk/ncc/Templates/ content_applications.aspx?NRMODE=Published&NRORIGINALURL=/Learning/Institutions/schoolsdir. htm?SchoolDetail=9287031ISpecial%20-%20Primary&NRNODEGUID={7BF3FCE4-28B4-49B8-8F4E-CEA308D6E556}& NRCACHEHINT=NoModifyGuest). Retrieved 8 August 2009.

[60] Northamptonshire County Council: Saturday Music and Performing Arts Centres (http://www.northamptonshire.gov.uk/en/
      councilservices/EducationandLearning/music/Pages/sat_centres.aspx). Retrieved 8 August 2009.
[61] Northamptonshire County Council: Music Service: Youth Groups (http://www.northamptonshire.gov.uk/en/councilservices/
      EducationandLearning/music/Pages/YouthGroups.aspx). Retrieved 8 August 2009.
[62] Tresham College: Our Campuses (http://www.tresham.ac.uk/about/our_campuses). Retrieved 8 August 2009.
[63] Tresham College: Our Courses (http://www.tresham.ac.uk/courses). Retrieved 8 August 2009.
[64] Tresham College: Higher Education (http://www.tresham.ac.uk/higher_education). Retrieved 8 August 2009.
[65] The University of Northampton: About Us (http://www.northampton.ac.uk/about/). Retrieved 8 August 2009.
[66] The University of Northampton: Course finder (http://www.northampton.ac.uk/courses/search/). Retrieved 8 August 2009.
[67] "New £4.2m Irthlingborough outpatients clinic opens" (http://www.bbc.co.uk/news/uk-england-northamptonshire-12379676). BBC
      News. 7 February 2011. . Retrieved 7th February 2011.
[68] Tite, Nick (2008-07-14). "Rabbit caused water contamination at Pitsford - Northants ET" (http://www.northantset.co.uk/news/
      Rabbit-caused-water-contamination-at.4286344.jp). Northants Evening Telegraph. . Retrieved 2008-08-22.
[69] "Sickness bug found in tap water" (http://news.bbc.co.uk/1/hi/england/northamptonshire/7472619.stm). BBC. 2008-06-25. .
      Retrieved 2008-07-15.
[70] "BBC News". *News at Ten, BBC One* (BBC). 2008-07-14.
[71] "Anglian Water" (http://www.anglianwater.co.uk/index.php?sectionid=51&parentid=50&contentid=1136), Press Release
[72] "Water crisis: All clear for tap water - and up to £30 compensation! - Northampton Chronicle and Echo" (http://www.northamptonchron.
      co.uk/news/25-each-compensation-for-water.4255069.jp). Chronicle & Echo. 2008-07-05. . Retrieved 2008-08-22.
[73] Britten, Nick (2009-02-23). "Corby station" (http://www.telegraph.co.uk/news/uknews/road-and-rail-transport/4787477/
      The-17million-train-station-with-only-one-service-a-day.html). London: Telegraph.co.uk. . Retrieved 2010-09-25.
[74] Network South East routes (http://www.nsers.org.uk/nse2.htm)
[75] "SMJR" (http://www.smjr.info). Smjr.info. 2010-09-19. . Retrieved 2010-09-25.
[76] (http://www.atoc.org/general/ConnectingCommunitiesReport_S10.pdf) ATOC Connecting Communities Report
[77] Network Rail: East Midlands Draft Route Utilisation Strategy (http://www.networkrail.co.uk/browse documents/rus documents/route
      utilisation strategies/east midlands/east midlands rus draft for consultation.pdf) Access date: 4th January 2010]
[78] Stagecoach Northants (http://www.stagecoachbus.com/northants/index.html)
[79] "First Northampton: Timetables" (http://www.firstgroup.com/ukbus/eastmidlands/northampton/timetables/index.php?operator=20&
      page=1&redirect=no). Firstgroup.com. 2010-09-19. . Retrieved 2010-09-25.
[80] http://www.ead.eurocontrol.int/eadbasic/pamslight-113B93979B7148E51E98BD22C3FF290E/7FE5QZZF3FXUS/EN/AIP/AD/
      EG_AD_2_EGBK_en_2011-03-10.pdf
[81] http://www.theroadmender.com/

References

- Greenall, R. L. (1979) *A History of Northamptonshire* Phillimore & Co. Ltd. ISBN 1-86077-147-5.

External links

- Northamptonshire County Council (http://www.northamptonshire.gov.uk/en/Pages/HomePage.aspx)
- Northamptonshire (http://www.dmoz.org/Regional/Europe/United_Kingdom/England/Northamptonshire/)
  at the Open Directory Project
- Northamptonshire Images and Information (http://www.northamptonshire.co.uk/)
- 1894/5 description (http://www.uk-genealogy.org.uk/gazetteer/england/Northamptonshire/)
- Northants Forum (http://northantscommunitycafe.co.uk/)
- Local Theatre in Northamptonshire (http://www.theatrenights.com/index.php?list=Events&location=4)
- Northamptonshire History Website (http://www.northamptonshire-history.org.uk/)
- Northamptonshire Tourism Website (http://www.explorenorthamptonshire.co.uk/)
- Northamptonshire Guide Website (http://www.northamptonshireguide.co.uk/)
- Visit Northamptonshire Website (http://www.visitnorthamptonshire.co.uk/)
- Northamptonshire Online Forum (http://www.northamptonshireforum.co.uk/)
- NorthantsSavings.com (Northamptonshire deals) (http://www.northantssavings.com/)

Corby

<table>
<tr><td colspan="2" align="center">Borough of Corby</td></tr>
<tr><td colspan="2" align="center">— Town & Borough —</td></tr>
<tr><td colspan="2">
Corby town centre skyline, seen from Oakley Woods</td></tr>
<tr><td colspan="2">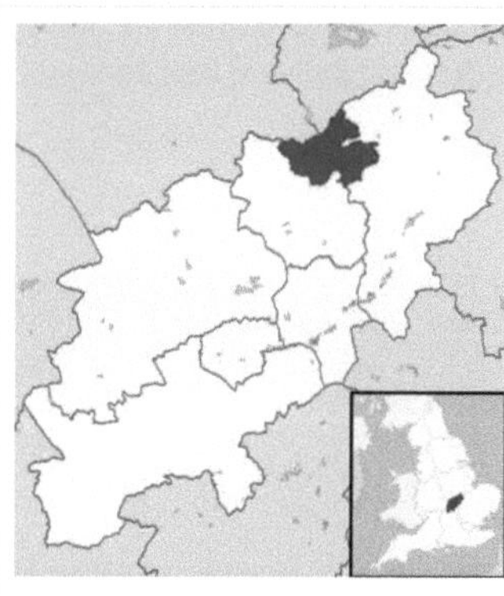
Borough of Corby shown within Northamptonshire</td></tr>
<tr><td colspan="2" align="center">Coordinates: 52°29′29″N 0°41′47″W</td></tr>
<tr><td>Sovereign state</td><td>United Kingdom</td></tr>
<tr><td>Constituent country</td><td>England</td></tr>
<tr><td>Region</td><td>East Midlands</td></tr>
<tr><td>Non-metropolitan county</td><td>Northamptonshire</td></tr>
<tr><td>Status</td><td>Non-metropolitan district</td></tr>
<tr><td>Admin HQ</td><td>Corby</td></tr>
<tr><td>Incorporated</td><td></td></tr>
<tr><td>Government</td><td></td></tr>
<tr><td>• Type</td><td>Non-metropolitan district council</td></tr>
<tr><td>• Borough council</td><td>Corby Borough Council (Labour)</td></tr>
<tr><td>• Mayor</td><td>Gail McDade</td></tr>
<tr><td>• MPs</td><td>Louise Mensch (Conservative)</td></tr>
<tr><td>Population (2010 est.)</td><td></td></tr>
<tr><td>• Total</td><td>55,800</td></tr>
</table>

• Rank	314th (of 326)
Time zone	GMT (UTC0)
• Summer (DST)	BST (UTC+1)
Postcodes	NN17-NN18
Area code(s)	01536
ONS code	34UB
OS grid reference	SP897887
Website	[www.corby.gov.uk www.corby.gov.uk]

Corby is borough located in the county of Northamptonshire. Corby Town is 23 miles north-east of the county town, Northampton. The borough had a population of 53,174 at the 2001 Census; the town on its own accounted for 49,222 of this figure. Figures released in March 2010 revealed that Corby has the fastest growing population in both Northamptonshire and the whole of England. The Borough of Corby borders onto the Borough of Kettering, the District of East Northamptonshire, the District of Harborough and the unitary authority county of Rutland. The town was at one time known locally as "Little Scotland" due to the large number of Scottish migrant workers who came to Corby for its steelworks.

Borough settlements

The Borough of Corby consists of the town of Corby, as well as the villages of Weldon, Rockingham, Gretton, Cottingham, Middleton, East Carlton, Stanion, and Little Stanion.

History

Early history

Mesolithic and Neolithic artefacts have been found in the area surrounding Corby and human remains dating to the Bronze Age were found in 1970 at Cowthick.[1] The first evidence of permanent settlement comes from the 8th century when Danish invaders arrived and the settlement became known as "Kori's by" – Kori's settlement. The settlement was recorded in the Domesday Book of 1086 as "Corbei". Corby's emblem, the raven, derives from an alternative meaning of this word. These Danish roots were recognised in the naming of the most southern of the town's housing estates, Danesholme, around which one of the Danish settlements was located.

Corby was granted the right to hold two annual fairs and a market by Henry III in 1226. In 1568 Corby was granted a charter by Elizabeth I that exempted local landowners from tolls (the fee paid by travellers to use the long distance public roads), dues (an early form of income tax)[2] and gave all men the right to refuse to serve in the local militia.[3] A popular legend is that the Queen was hunting in Rockingham Forest when she (dependent on the legend) either fell from her horse or became trapped in a bog whilst riding. Upon being rescued by villagers from Corby she granted the charter in gratitude for her rescue. Another popular explanation is that it was granted as a favour to her alleged lover Sir Christopher Hatton.

The Corby Pole Fair is an event that has taken place every 20 years since 1862 in celebration of the charter. The next pole fair is to be held in 2022.

From rural village to industrial town

The local area has been worked for iron ore since Roman times. An ironstone industry developed in the 19th century with the coming of the railways and the discovery of extensive ironstone beds. By 1910 an ironstone works had been established. In 1931 Corby was a small village with a population of around 1,500. It grew rapidly into a reasonably sized industrial town, when the owners of the ironstone works, the steel firm Stewarts & Lloyds, decided to build a large integrated ironstone and steel works on the site. The start of construction in 1934 drew workers from all over the country including many workers from the depressed West of Scotland and Irish labourers. The first steel was produced in October 1935 and for decades afterwards the steel works dominated the town. By 1939 the population had grown to around 12,000, at which time Corby was thought to be the largest "village" in the country, but it was at that point that Corby was re-designated an urban district (see the Local Government section below).

The 1940s and 1950s

During World War II the Corby steelworks were expected to be a target for German bombers but in the event there were only a few bombs dropped by solitary planes and there were no casualties. This may be because the whole area was blanketed in huge dense black, low lying clouds created artificially by the intentional burning of oil and latex to hide the glowing Bessemer converter furnaces at the steelworks from German bomber crews.[4] The only known remaining scars from German attacks can be found in the form of bullet holes visible on the front fascia of the old post office in Corby village (now known as Maddison's Bar and Storm nightclub). Nobody really knows the exact circumstances under which the attack occurred, but a local apocryphal tale tells of a lone pilot making his way back to Germany after a successful raid on Coventry who spotted some lights so decided to finish off his already depleted stock of bullets. Sadly, the authenticity of this romanticised tale can neither be verified or denied, but it is certainly the most popular theory among locals. The Corby steelworks made a notable contribution to the war effort by manufacturing the steel tubes used in Operation Pluto (Pipe Line Under the Ocean) to supply fuel to Allied forces on the European continent.

In 1950, with a population of 18,000, Corby was designated a New Town with William Holford as its architect. By 1951, he prepared the development plan with a car-friendly layout and many areas of open space and woodland. In 1952, Holford produced the town centre plan and in 1954 the layout for the first 500 houses.[5] The town now underwent its second wave of expansion, mainly from Scotland.

The decline of the steel industry

Sundew dragline excavator was a local landmark

In 1967 the British steel industry was nationalised and the Stewarts & Lloyds steel tube works at Corby became part of British Steel. In 1973 the government approved a strategy of consolidating steel making in five main areas – South Wales, Sheffield, Scunthorpe, Teesside and Scotland – most of which are coastal sites with access to economic supplies of iron rich imported ores. Thus in 1975 the government agreed a programme that would lead to the phasing-out of steel making in Corby.[6] In November 1979 the end of iron and steel making in Corby was formally announced. By the end of 1981 over 5,000 jobs had been lost from British Steel in Corby, and further cuts took the total loss to 11,000 jobs, leading to an unemployment rate of over 30%.[7] [8] Steel tube making continued, however, initially being supplied with steel by rail from Teesside and later from South Wales.

Redevelopment

New industry was subsequently attracted to the town and by 1991 unemployment had returned to the national average.[9] The recovery of Corby was explained in 1990 by John Redwood, then a junior minister in the Department of Trade and Industry, as being a result of the establishment of an Enterprise Zone, the promotion of Corby by the government, the work of private investors and the skills of the work force. Others believe the town's recovery was significantly assisted by its central location and substantial grants from the EU.[10]

Corby's CCGT power station

To the north of Corby, on the industrial estates, is a 350MW power station built in 1994; and the Rockingham Motor Speedway built in 2001.

Politics

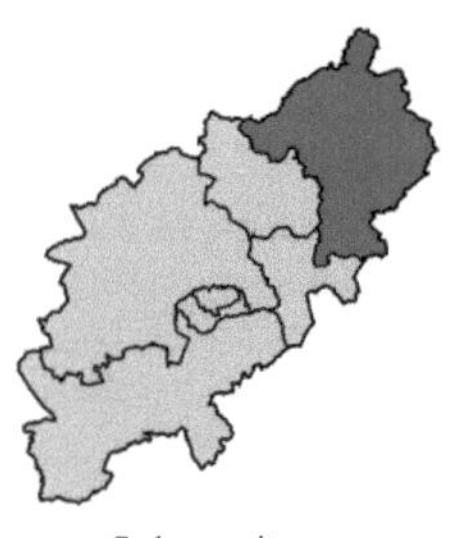
Corby constituency

The Corby constituency contains parts of traditionally Conservative East Northamptonshire that balance the traditionally Labour town of Corby leading to a marginal constituency that has gone to the party forming the national government in every general election since the creation of the constituency in 1983. In the 2005 General Election, Labour won Corby by a majority of just over 1,000. At the 2010 General Election, the sitting Labour MP Phil Hope lost this seat to Louise Bagshawe, the Conservative Party candidate, who is now Louise Mensch after her marriage. Corby Borough Council has been controlled by the Labour party since 1979. In 2007 the council had 16 Labour representatives, eight Conservatives and five Liberal Democrats.

Elections

- Corby Borough Council Elections 2007
- European Parliament Elections 2004 (East Midlands Constituency)
- United Kingdom General Election 2005 (Corby Constituency)

Society and culture

Scottish migration to Corby has created a unique population in the borough, evidenced most clearly in the 'Corby accent', referred to as 'Corbyite', which is often described as sounding Glaswegian. The link with Scotland is a strong feature of the area: according to the 2001 Census, there were 10,063 Scottish-born in the Corby Urban Area – 18.9% of the population. A further 1.3 per cent were born in Northern Ireland. It has been estimated that a further third of the population are Scottish or of Scottish descent.[11]

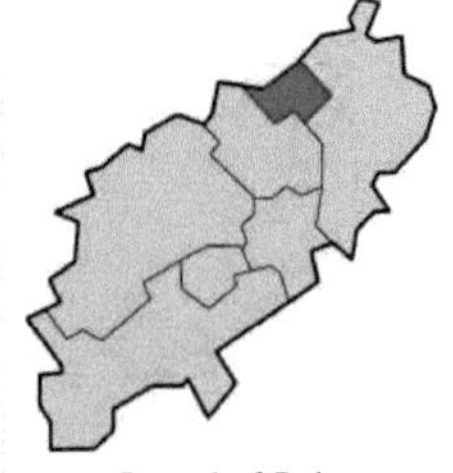
Borough of Corby

The Scottish heritage is cherished by many inhabitants. There are Scottish social and sporting clubs and there are many fervent supporters of the Rangers and Celtic football clubs (indeed, Corby is home to the largest Rangers Supporters' Club outside of Glasgow and Northern Ireland). Many shops sell Scottish foods and a supermarket even introduced Gaelic signs[12] to their Corby store (but they have since removed them). An annual Highland Gathering featuring traditional Scottish music and dancing is held in the town. Corby is the only town in England apart from London with two Church of Scotland churches.[13]

Corby's boating lake

The song *Steeltown* by Big Country (title track of the album) was written about the town of Corby, telling how many Scots went to work there, but found themselves unemployed when the steelworks declined. (Source: Melody Maker, 1984)

According to the 2001 Census 1.7% of the population are non-white and the average age of the population (37.2) is slightly lower than the average for England and Wales (38.6).

November 2010 saw the opening of the Corby Cube, a major development in the town centre. As well as new council chamber, registrar office, and public library, the Cube is home to a 450 seat theatre and 100 capacity studio theatre. A programme of live theatre, dance, music and standup comedy is complemented by a participation programme encouraging all parts of Corby community to get involved. Recently the theatre started screening films, twice a week and including current mainstream releases and the best in world, independent and art house cinema.

A crater on Mars discovered in the late 1970's was named after Corby, in reference to a famous transcript of a conversation in June 1969 between the crew of the Apollo 11 mission and mission control, whereby world news was relayed to the crew, amongst it was the news that "in Corby, an Irishman named John Coil won the World's Porridge Eating Championship by consuming 23 bowls of instant oatmeal in 10 minutes".The reply from Apollo 11 "I'd like to enter Aldrin in the porridge eating contest next time; he's on the 19th bowl. Rodger."

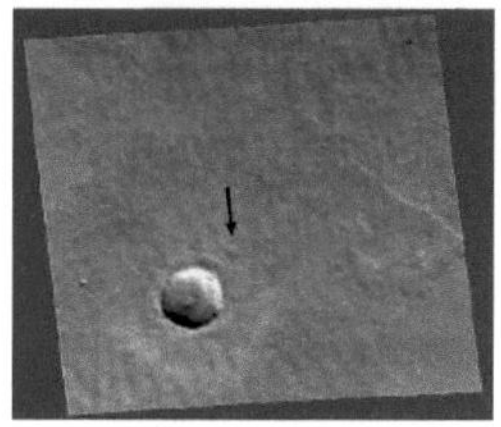
The Corby Crater

Sport & Leisure

Corby has two Non-League football teams Corby Town F.C. who play at Steel Park and Stewarts & Lloyds Corby F.C. who play at Recreation Ground on Occupation Road.

Transport

Roads

The town is located along the A43, A427, A6003 and is six miles from the A14 at Kettering. Corby lies within two hours' drive of four international airports: Birmingham, Luton, Stansted and East Midlands. Being a new town, Corby's road network is different to that of older towns. There are several dual carriageways, most of the principal roads have wide reservations and high speed limits and pedestrian crossings over them are often underpasses. However, Corby is only connected by dual carriageway to one neighbouring town, Kettering (the A6003). All other roads into the town are single carriageways.

Buses

Corby is served by six bus routes under the Corby Star brand name.[14] The route X4 connects the town with Northampton, Milton Keynes, Peterborough and is also operated by Stagecoach[15] and Glasgow by National Express. Plans to build a new bus station in Corby are being considered by the council following the closure of the old bus station in August 2002.

Rail

Corby rail-link bus

Work begins on the railway station in March 2008

Corby used to be described as the largest town in the UK not to have a railway station, or access within five miles of one - many other towns and localities claim this dubious honour (e.g. Gosport, Dudley, Newcastle under Lyme) but all of these places have locally available railway stations in adjacent suburban areas. Previously the nearest station was seven miles south in Kettering since the closure of the original station under the Beeching Axe in April 1966. The Kettering-Corby-Melton Mowbray section remained open for freight and through passenger trains, passing through the 1,920 yard (1,756 metre) Corby Tunnel and crossing the River Welland on the colossal 82-arch Welland Viaduct.

The newly built station opened on 23 February 2009. East Midlands Trains runs hourly services to London St Pancras via Kettering and Wellingborough. There is also a limited peak time service running north to Oakham and Melton Mowbray.[16] It was later agreed that the 'Corby Rail Bus', the X1 service between Corby and Kettering operated by Stagecoach, would be kept running after the opening of the new station – though passenger numbers will be monitored and if they fall off significantly then the service may be reduced or terminated. Train services had been due to start on 14 December 2008, but EMT admitted that it failed to secure the four new trains it needed. An article in Corby's local newspaper stated that the service would be starting on 23 February 2009.[17] [18]

Corby railway station opened on 23 February 2009

Employment and education

Employment

Since the 1980s the unemployment rate has returned to a level closer to the national average (2.7% in October 2005).[19] Employment is biased towards manufacturing (36.8% compared with a regional average of 18.5%) and against public administration, health and education (10.0% compared with the regional average of 25.9%).[20] Much of industry is concentrated in purpose-built industrial estates on the outskirts of the town. Fairline Boats are manufactured here. Weetabix Limited make Weetos in the north of the town. RS Components are based near Rockingham Speedway.

Demographics

According to the 2001 Census the proportion of the working age population with degree-level qualifications (8.5%) is the lowest of all areas in England and Wales. 39.3% have no GCSE-equivalent qualifications at all.[21] The borough of Corby has the highest rate of teenage pregnancy in the East Midlands, outside of the metropolitan boroughs (unitary authorities), although Lincoln is very similar.

Schools

Brooke Weston Academy

The Corby campus of Tresham Institute of Further and Higher Education provides a range of vocational courses for post-16 students and adult learners. The nearest universities are the University of Northampton, 37 km (**unknown operator: u'strong'** mi) to the south and both the University of Leicester and De Montfort University in Leicester, 40 km (**unknown operator: u'strong'** mi) to the west.

Brooke Weston College, was one of only fifteen CTCs in England, opened in 1990. Brooke Weston CTC consistently achieved examination results in the top 5% of English state schools, and has been a City Academy since September 2008[22] after which it continued with those good results.

Lodge Park Technology College on Shetland Way

Since 1990 several of Corby's other secondary schools have fared less well with a series of poor examination results and critical inspection reports leading to mergers and closures, the most recent being the closure of Our Lady and Pope John School in 2005. Currently there are four secondary schools in Corby: Brooke Weston Academy, Lodge Park Technology College, Corby Business Academy (formerly Corby Community College) and The Kingswood School (A Specialist Arts College). Corby Community College has a special unit for children with severe special educational needs. All four schools have sixth forms for post-16 students.

Corby has 17 primary schools, of which two are Church of England schools, three are Roman Catholic and one for children with severe behavioural and emotional difficulties.

Regeneration and redevelopment

Corby has an Urban Regeneration Company - North Northants Development Company, which now covers the whole of North Northamptonshire rather than just Corby (it was previously known as Catalyst Corby). The company is working closely with Corby Borough Council, Land Securities (town centre owners), the East Midlands Development Agency and the Homes and Communities Agency to regenerate the town centre as part of the masterplan for the whole town. The population of Corby town is expected to double in the next 30 years through housing on large estates such as Prior's Hall, Little Stanion, Oakley Vale and Great Oakley.

In October 2007 Corby's new shopping precinct, Willow Place, opened.[23] In addition Parkland Gateway, the Borough's £50m investment situated adjacent to Willow Place and including a new Olympic-sized swimming pool and civic hub (due for completion in November 2010), is being built following its approval in January 2007. Work began on the project in October 2007 and the Corby East Midlands International Pool was officially opened by Olympian Mark Foster in July 2009. Although the Evolution Corby project is currently on hold, limited aesthetic augmentation work within the town centre continues.

Stephen Fry has previously been employed to do the voice-over work for a campaign running in London to entice people to move to Corby. The campaign is centred around advertisements in newspapers, on the London Underground and on local radio. An example of one of the posters in the 'More for your Money' campaign (photographed on the London Underground). A new campaign to persuade Londoners to move to North Northamptonshire, dubbed 'North Londonshire', is due to launch in the coming weeks.

Toxic waste contamination

In July 2009 Corby Borough Council was found liable for negligently exposing pregnant women to toxic waste during the reclamation of the former British Steel steelworks, causing birth defects to their children.[24] The judge found in favour of 16 of the 18 claimants, the oldest of whom was 22 at the time of the ruling. The ruling was significant as it was the first in the world to find that airborne pollution could cause such birth defects==Geography==

Climate

As with the rest of the British Isles and Northamptonshire, Corby experiences a maritime climate with cool summers and mild winters. The nearest official Metoffice weather station for which online records are available is Caldecott, about 3 miles north-north west of Corby town centre and in the Welland valley on the Northamptonshire-Leicestershire border. Observations have now ceased, and the nearest operational weather station is Wittering, some 13 miles to the north east. The exact situation of the weather station at Caldecott is at 53m and in a valley location, meaning generally higher daytime temperatures than Corby itself (at 100-130m altitude) but also higher frost averages. Caldecott is in fact frostier than the frost hollow at Shawbury (70.2 frost days vs. 64 frost days at Shawbury), where England's lowest December temperature was recorded. However, given temperatures between the 1961-90 period (displayed below) and 1971-2000 period have typically increased by 0.5 to 0.7 Celsius, offset by the fall in temperature of a similar value between the 53m of the weather station and the 120 metres or so of Corby town centre, the figures should provide accurate representation (in daytime values at least) of the town proper for the 1971-2000 period.

The lowest temperature recorded was −23.3 °C (**−unknown operator: u'strong'** °F) during January 1987,[25] the highest temperature of 35.0 °C (**unknown operator: u'strong'** °F) was recorded in August 1990.[26]

Twin towns

Corby is twinned with:

- ▨ Shijiazhuang, Hebei, China

References

[1] An Archaeological Resource Assessment of the Neolithic and Bronze Age in Northamptonshire (http://www.le.ac.uk/archaeology/research/projects/ eastmidsfw/pdfs/14nhneba.pdf)

[2] Corby Borough Council - The History of Corby (http://www.corby.gov.uk/an/ wc.exe/ao2/View/?Doc=12317&Site=1182)

[3] Corby Pole Fair Charter (http://www.bbc.co.uk/northamptonshire/asop/corby/ pole_fair/pole_charter.shtml)

[4] Memories of the Second World War (http://www.bbc.co.uk/dna/ww2/ A2792072)

[5] Mervyn Miller, William Graham, Baron Holford (1907-1975) (http://www. oxforddnb.com/view/article/31245|"Holford,), Oxford Dictionary of National Biography, Oxford University Press 2004; accessed 21 Jan 2012

[6] History of British Steel (http://www.corusgroup.com/file_source/StaticFiles/ Corporate/History_BS.pdf)

[7] Memorandum by Corby Borough Council (NT 50) (http://www.parliament.the-stationery-office.co.uk/pa/cm200102/cmselect/cmtlgr/ 603/603ap33.htm)

[8] The State of the Regions (http://www.lgiu.gov.uk/admin/images/uploaded/Stateofregions.pdf), Local Government Information Unit

Rockingham Motor Speedway

[9] Corby Northamptonshire through time - Historical Statistics on Work and Poverty (http://www.visionofbritain.org.uk/data_rate_page.
     jsp?u_id=10189020&c_id=10001043&data_theme=T_WK&id=0)
[10] (http://www.catalyst-corby.co.uk/doingbusiness/sectors_distribution.php) Corby is already recognised as a prime location for
     distribution and logistics
[11] The English town that's truly Scottish (http://heritage.scotsman.com/people.cfm?id=2089062005)
[12] Gaelic welcome in store (http://news.bbc.co.uk/1/hi/england/2858327.stm)
[13] Church of Scotland - Presbytery of England (http://www.churchofscotland.org.uk/contact/contactmap47.htm)
[14] Stagecoach Northants: Corby Star network (http://www.stagecoachbus.com/uploads/corbymapdownloadlo.pdf) Accessed 9 April 2010
[15] Stagecoach X4 (http://www.stagecoachbus.com/northamptonshire-X4.aspx) Accessed 9 April 2010
[16] New Service to run north to Oakham (http://www.northantset.co.uk/corby/Corby-rail-users39-joy-at.4433720.jp)
[17] Corby article (http://www.northantset.co.uk/corby/Train-service-will-leave-from.4959560.jp)
[18] Corby train delays labelled 'shambolic' (http://www.northantset.co.uk/corby/Corby-train-delays-labelled-39shambolic39.4675913.jp)
[19] Geographical Statistical Information - Unemployment (http://www.go-em.gov.uk/geography-skin.php?set=unemploy&LA=34UB)
[20] Geographical Statistical Information (http://www.go-em.gov.uk/geography-skin.php?LA=34UB&x=0&county=northants&y=1)
[21] Census 2001 (http://www.statistics.gov.uk/census2001/profiles/34UB-A.asp)
[22] Full list of academies (http://politics.guardian.co.uk/publicservices/story/0,,1753474,00.html)
[23] Mayor declares Willow Place officially open (http://www.nndev.co.uk/newsdetail.php?nid=23)
[24] Williams, Rachel. Council found liable for children's exposure to toxic waste (http://www.guardian.co.uk/society/2009/jul/29/
     corby-council-steelworks-disabilities). 29 July 2009. Guardian.co.uk. Retrieved 29 July 2009.
[25] "1987 cold" (http://www.personal.dundee.ac.uk/~taharley/coldest_days.htm). TA Harley British Weather. .
[26] "1990 Heatwave" (http://archive.greenpeace.org/climate/database/records/zgpz0697.html). Greenpeace. .

External links

- Corby Local Shops News & Events (http://www.corbylocal.com/)
- More in Corby - place marketing site (http://www.moreincorby.co.uk/)
- Corby Borough Council (http://www.corby.gov.uk)
- North Northants Development Company (http://www.nndev.co.uk/)
- Local news (http://www.corbytoday.co.uk/)
- Corby Radio (http://www.corbyradio.com/)
- BBC website about Corby (http://www.bbc.co.uk/legacies/immig_emig/
  england/northants/article_1.shtml)

Phoenix Parkway with the power
station in the distance

- The English town that's truly Scottish (http://heritage.scotsman.com/people.cfm?id=2089062005)
  (Scotsman.com)
- WonderWorld, Corby, UK (http://h2g2.com/dna/h2g2/A787809) on h2g2
- Corby Crater website (http://www.fourth-millennium.net/space-exploration/space-exploration-corby-crater.
  html)
- Corby Satellite map and weather (http://rumbletum.org/Europe/United+Kingdom/United+Kingdom+
  (general)/_2652381_Corby.html)
- Rockingham Racetrack (http://www.rockingham.co.uk)
- Corby Skyscrapers! (http://www.skyscrapercity.com/showthread.php?t=406630)
- Corby pictures and information (http://www.corby.me.uk/)
- Corby Northern Orbital Road (http://www.corbyorbital.com/)

St._Mary_the_Virgin,_Brampton_Ash

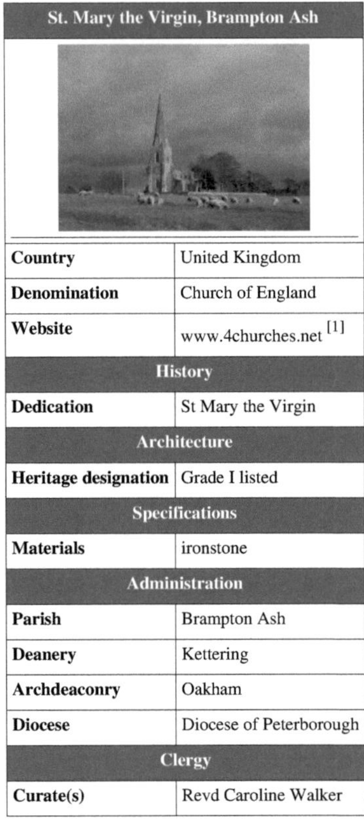

St. Mary the Virgin, Brampton Ash	
Country	United Kingdom
Denomination	Church of England
Website	www.4churches.net [1]
History	
Dedication	St Mary the Virgin
Architecture	
Heritage designation	Grade I listed
Specifications	
Materials	ironstone
Administration	
Parish	Brampton Ash
Deanery	Kettering
Archdeaconry	Oakham
Diocese	Diocese of Peterborough
Clergy	
Curate(s)	Revd Caroline Walker

St. Mary the Virgin is the local parish Church of England church for Brampton Ash, Northamptonshire. Sitting in the Diocese of Peterborough, the ironstone church boasts fine carvings of lions.

It is well lit at night, and can be seen across the Welland Valley for miles around.

The church is largely 13th and 14th century with some restoration in the 19th century.

Memorials

On the south wall of the chancel, Charles Norwich, died 1605, and wife, 2 kneeling figures under arch.

On the north wall of the chancel, Thomas Farmer, died 1764, and 2 other tablets.

On the west wall of the south aisle: George Bosworth, died 1804; marble tablet with 2 weeping willows bending over an urn. 2 19th century tablets alongside.[2]

List of Rectors

- Thomas , 1230
- John , ?
- Richard de Flammevill , 1264
- Hugh de Valle , 1294
- John de Cumpton , 1319
- John de Stamford , 1325
- John de Felmersham , 1343
- John de Kirkham , 1345
- Oliver de Dineley , 1347
- William de Gayrstang , 1348
- John de Totyngton , 1349
- Robert Wylmot , 1350
- Henry de Greynesby , 1352
- John Noioun , 1352
- John Essex , 1356
- William de Cabernaco , 1357
- William Robert , 1358
- Peter la Sudria , 1358
- John Wade , 1366
- John Millicent , 1375
- John Rodyngton , 1379
- Thomas Ilneston , 1382
- John Wade , 1386
- Walter Tyngyn , ?
- John Bottlesnam , 1395
- William Topclyff , 1396
- William Hole , 1403
- William Fraunceys , ?
- William Islip , 1410
- William Maidwell , 1412
- John Fynche , 1420
- William Halle , 1424
- William Newbery , 1429
- Thomas Webster , 1440
- William Dene , 1441
- Thomas Smyth , 1443
- John Mallory , 1443
- John Smyth , 1446
- John Mallory , 1451

- Richard Hyndeman , 1452
- John Jonys , 1459
- Richard Spicer , 1463
- Walter Oudeby , 1466
- John Bigcrofte , 1483
- David Barker , 1503
- John Devyas , 1509
- Edmund Olyver , 1546
- Anthony Palmer , 1573
- Andrew Broughton , 1577
- William Addison , 1615
- Richard Cumberland , 1662
- Samuel Blackwell , 1691
- Robert Browne , 1720
- Philip Bliss , 1733
- Thomas Farrer , 1734
- William Arden , 1764
- Samuel Rogers , 1769
- Samuel Heyrick , 1790
- Hon. Charles Dundas , 1841
- Sidney Lidderdale Smith , 1844
- Austin Ainsworth Slack , 1904
- Thomas Beckenn Avening Saunders , 1908
- Gerard Cokayne Vecqueray , 1910

In 1928, the benefice was united with Dingley, and the incumbent ceased to be resident at Brampton Ash.

References

[1] http://www.4churches.net
[2] Buildings of England: Northamptonshire: Nikolaus Pevsner. p119

A427_road

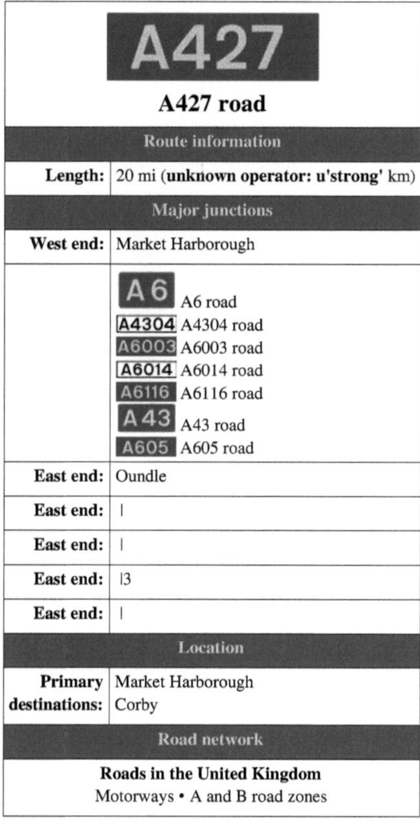

A427 road

Route information	
Length:	20 mi (**unknown operator: u'strong'** km)
Major junctions	
West end:	Market Harborough
	A6 A6 road A4304 A4304 road A6003 A6003 road A6014 A6014 road A6116 A6116 road A43 A43 road A605 A605 road
East end:	Oundle
East end:	I
East end:	I
East end:	I3
East end:	I
Location	
Primary destinations:	Market Harborough Corby
Road network	
Roads in the United Kingdom Motorways • A and B road zones	

The **A427 road** is a major road in the English Midlands. It connects the Leicestershire town of Market Harborough and the A6 with the Northamptonshire town of Oundle and the A605.

History

When first designated the A427 ran from a junction with the A45 southwest of Rugby through Rugby itself to Clifton upon Dunsmore crossing the A5 through Catthorpe (past what is now the Catthorpe Interchange between the M1 motorway, the M6 and the A14 road). It then passed through Swinford, South Kilworth, North Kilworth, Husbands Bosworth, Theddingworth, Lubenham and Market Harborough. From there along its current route to Oundle. By 1976 the former A4114 road from Coventry via Lutterworth to North Kilworth had taken the designation A427 in place of the earlier route. By 2010 the A427 ended in Market Harborough.

A section of the original route SW of Rugby is now the A4071. From Rugby to South Kilworth it is the B5414, from South Kilworth and North Kilworth is no longer classified. The section from North Kilworth and Market Harborough is now part of the A4304. The 1976 route has now taken the B4027 and A4303 designation between Coventry and Market Harborough.

Running through Corby, the A427 meets with the A6003 on the western edge of the town and the A43 in the Weldon North Industrial Estate, where it is briefly dual carriageway.

Route

From West to East its route is:

* Market Harborough
* Dingley
* Brampton Ash
* Stoke Albany (bypassed)
* Wilbarston (bypassed)
* Corby (where it is crossed by the A43)
* Weldon
* Upper Benefield
* Lower Benefield
* Oundle

Desborough

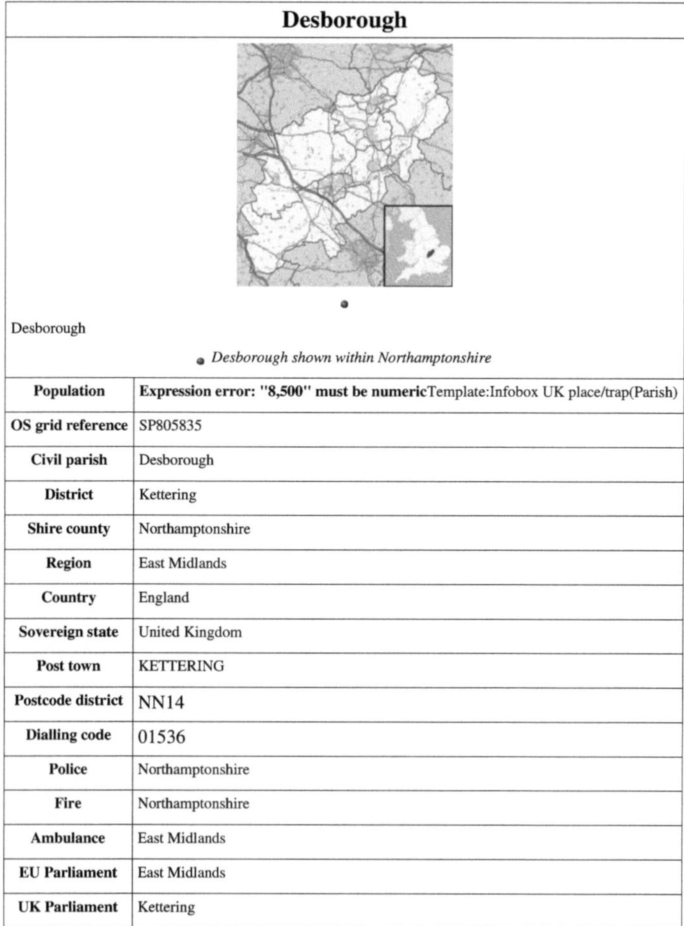

<table>
<tr><th colspan="2" align="center">Desborough</th></tr>
<tr><td colspan="2" align="center">Desborough

Desborough shown within Northamptonshire</td></tr>
<tr><td>Population</td><td>Expression error: "8,500" must be numericTemplate:Infobox UK place/trap(Parish)</td></tr>
<tr><td>OS grid reference</td><td>SP805835</td></tr>
<tr><td>Civil parish</td><td>Desborough</td></tr>
<tr><td>District</td><td>Kettering</td></tr>
<tr><td>Shire county</td><td>Northamptonshire</td></tr>
<tr><td>Region</td><td>East Midlands</td></tr>
<tr><td>Country</td><td>England</td></tr>
<tr><td>Sovereign state</td><td>United Kingdom</td></tr>
<tr><td>Post town</td><td>KETTERING</td></tr>
<tr><td>Postcode district</td><td>NN14</td></tr>
<tr><td>Dialling code</td><td>01536</td></tr>
<tr><td>Police</td><td>Northamptonshire</td></tr>
<tr><td>Fire</td><td>Northamptonshire</td></tr>
<tr><td>Ambulance</td><td>East Midlands</td></tr>
<tr><td>EU Parliament</td><td>East Midlands</td></tr>
<tr><td>UK Parliament</td><td>Kettering</td></tr>
</table>

Desborough is a town in Northamptonshire, England. It is located in the Ise Valley, between Market Harborough and Kettering. In the 19th century, the town was an industrial centre for weaving and shoe making. It has a long association with the Co-operative movement.[1] Modern Desborough is a residential centre, with new homes and industry being developed to the north of the old town centre.

Overview

Desborough developed around the spinning and weaving industries, by the nineteenth century specialising in silk. Many archaeological finds from the Celtic and Anglo-Saxon periods have been made in the town, some of which have been inducted into the collection of London's British Museum. The Desborough Mirror is an example of this. [2]

Desborough is 8 km (**unknown operator: u'strong'** mi) south-east of Market Harborough, 8 km (**unknown operator: u'strong'** mi) north-west of Kettering and 8 km (**unknown operator: u'strong'** mi) south-west of Corby.

The Kettering Leg of the Student cross pilgrimage leaves from near Desborough every year.

The A6 Rothwell-Desborough Bypass opened on August 14, 2003.

Notable buildings in the town include the thirteenth century parish church. The Domesday Book of 1086 refers to Desborough, in modern day Northamptonshire, as a 'place of judgement'. In fact the name itself is thought to have derived from 'Disburg' which meant a sacred and fortified place.

In the High Street, as a centrepiece of what is now the Market Square, stands a pillar. Locals call it the Town Cross, despite it being a square column with a stone ball on top. It is perhaps better referred to as an obelisk. Its origins are thought to be a gateway pillar from Harrington Hall.

Desborough's origins lie in the Bronze Age some 2000 years BC. Urns from this period have been found in and around the town. The most important archeological find was the 1st Century Desborough Mirror which is now in the British Museum as is an Anglo-Saxon necklace found in the Paddock Lane area of the town which comprises gold beads, a gold cross and a red garnet. Other stone artifacts are on display in the parish church of St Giles.

The Cross, Desborough.

St Giles Church

St Giles Church is the oldest surviving building in the town having been built in about 1225 AD. It is believed to stand on the site of an earlier Saxon church. Relics of the town's history including part of an Anglo-Saxon cross carved from stone, a Tudor rood screen and reminders of the Civil War. Close by the church is the 18th century Church House the 19th century Desborough House with its stucco and Doric pillars, now the Services Club.

On 7 September 1969 the Anglican (Church of England) and Methodist partnership was inaugurated in the presence of the Bishop of Peterborough and the Chairman of the Oxford District. Since that time a Methodist minister has been working in equal partnership with the Anglican vicar. St Giles is part of the United Benefice of Desborough and Brampton Ash with Braybrooke and Dingley. [3]

St Giles has regular church festivals including one of the UKs longest running (since 1998) and largest (over 100 trees) Christmas Tree Festival. The trees are contributed by local organisations, companies, individuals and families.

In addition to the parish church there is a Baptist church, [4] a United Reformed Church and the Roman Catholic Church of the Holy Trinity.

The Old Manor House

The Old Manor House in Gold Street retains many features of its late 17th century origins. Ferdinando Poulton, a Roman Catholic lawyer was Lord of the Manor, reputedly one of the Gunpowder Plot conspirators.

Government and community

Desborough comes under Northamptonshire County Council and Kettering Borough Council, as well as having its own Town Council [5].

It is one of the founding 12 members of the Charter of European Rural Communities[6] and through this has links with 26 other EU member towns and villages. Through the Desborough and District Twinning Association[7] the town is twinned with Neuville de Poitou in the Vienne departement of France and with Bievre in Belgium.

The Desborough Community Development Trust campaigns for improvements to the town.

Employment

In the 17th century the town developed as a spinning and weaving centre. Using local wool and flax, the town's factories produced fine cloth and linen until the mid 19th century. Silk weaving then developed in a Paddock Lane factory.

To counter exploitation by agents and employers, local men founded the Desborough Co-operative Society in 1863. Starting with local shops and then a corset and lingerie factory, the Desborough Co-op still has a department store, a bank, a supermarket, a travel agents, a ladies shoe and clothing shop and a couple of corner stores.

The former Co-op Corset Factory is now owned by Eveden Ltd[8] making lingerie and swimwear. The site includes the original Victorian factory and, immediately opposite, Eveden's warehousing and UK factory shop.

The former Co-operative Society Sports Ground with its football field and tennis courts is now the site of a major housing development called Desbeau Park. Desbeau was the name of one of the range of lingerie made at the Corset Factory.

The Desborough Co-op was recently purchased by the larger East Midlands Co-op who have since closed the bank and the ladies shoe and clothing shop. They have also turned one of the corner shops in to a funeral directors.

The former RS Lawrence's shoe factory site on the High Street was sold by the Midland Co-operative Society to Kettering Borough Council.

Sport & Leisure

Desborough has a Non-League football team Desborough Town F.C. who play at Waterworks Field.

Schools

There is one primary school and one infant/junior school in Desborough, Loatlands Primary School,[9] Havelock Infants,[10] and Havelock Junior[11] but no secondary school so children aged 11–16 mostly attend Montsaye Community College in Rothwell.

New development

Several housing projects are currently taking place, such as "The Grange", a joint project by several house building companies, and Jelson Homes's "Carisbrooke Grange [12]".

Magnetic Park is a project to build a business park in Desborough, most construction has been completed.[13]

There were plans to build a Sainsburys near Magnetic Park but Kettering Borough Council's planning committee turned down the company's proposals. Committee members voted 4-3, with one abstention, against the plans.[14] On 25 January 2012, the Council again rejected the plans by 7 to 1 votes, with one abstention.[15]

Plans have been submitted to Kettering to redevelop the town centre Lawrence Factory site with a new Tesco store. In September 2011, Kettering published a report recommending approval of the Tesco plans, but the Planning Committee [16] due to consider the application on 13 September 2011 was cancelled. On 25 January 2012, Kettering Borough Council voted unanimously to approve the plans.[17]

References

[1] http://www.afamilystory.co.uk/desborough/cooperative/jubilee-souvenir.aspx

[2] http://www.desboroughheritagecentre.co.uk/about-desborough-heritage.htm

[3] 4churches.net (http://www.4churches.net)

[4] http://www.desboroughbaptist.org.uk

[5] http://www.desboroughtowncouncil.gov.uk/Core/Desborough-Tc/Pages/Desborough_Town_Council_1.aspx

[6] European Charter (http://www.europeancharter.eu)

[7] Twinning Home Page (http://desboroughtown.co.uk/index.php/twinning)

[8] Eveden (http://www.eveden.com)

[9] Loatlands School Web Site (http://e-voice.org.uk/loatlandsprimaryonline/)

[10] Havelock Infant School Web Site (http://www.havelockinfant.ik.org/)

[11] Havelock Junior School (http://www.havelockjunior-v2.ik.org/)

[12] http://www.jelson.co.uk/sites/desborough/default.htm

[13] http://www.estatesgazette.com/propertylink/advert/desborough_kettering-desborough_kettering-3282438.htm

[14] Sainsburys-desborough.co.uk (http://www.sainsburys-desborough.co.uk)

[15] http://www.northantset.co.uk/community/local-information/supermarket_approved_but_saga_will_continue_1_3454295

[16] http://www.kettering.gov.uk/site/scripts/meetings_info.php?meetingID=915

[17] http://www.northantset.co.uk/community/local-information/supermarket_approved_but_saga_will_continue_1_3454295

(http://www.lawoodcraft.co.uk)>LA Woodcraft-Specialist in French Polishing and all types of woodcrafts and restoration.

External links

- Desborough & District Twinning Association & Charter (http://www.desboroughtown.co.uk)

Kettering

<table>
<tr><th colspan="2">Kettering</th></tr>
<tr><td colspan="2" align="center">
Part of pedestrianised Kettering High Street</td></tr>
<tr><td colspan="2" align="center">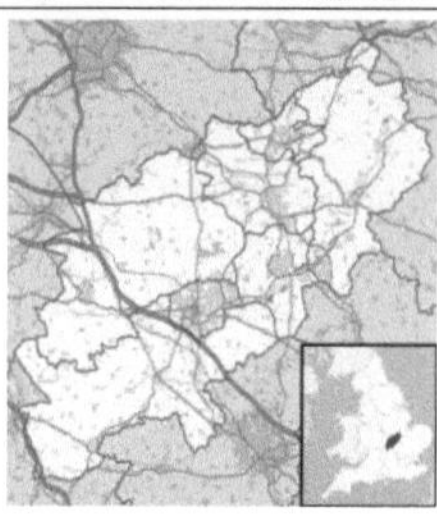
Kettering

Kettering shown within Northamptonshire</td></tr>
<tr><td>Population</td><td>Expression error: "51,063" must be numericTemplate:Infobox UK place/trap[1] (2001 Census)</td></tr>
<tr><td>OS grid reference</td><td>SP8778</td></tr>
<tr><td>- London</td><td>81 miles (unknown operator: u'strong' km)</td></tr>
<tr><td>District</td><td>Kettering</td></tr>
<tr><td>Shire county</td><td>Northamptonshire</td></tr>
<tr><td>Region</td><td>East Midlands</td></tr>
<tr><td>Country</td><td>England</td></tr>
<tr><td>Sovereign state</td><td>United Kingdom</td></tr>
<tr><td>Post town</td><td>KETTERING</td></tr>
<tr><td>Postcode district</td><td>NN15, NN16</td></tr>
<tr><td>Dialling code</td><td>01536</td></tr>
<tr><td>Police</td><td>Northamptonshire</td></tr>
<tr><td>Fire</td><td>Northamptonshire</td></tr>
<tr><td>Ambulance</td><td>East Midlands</td></tr>
<tr><td>EU Parliament</td><td>East Midlands</td></tr>
<tr><td>UK Parliament</td><td>Kettering</td></tr>
<tr><td>Website</td><td>http://www.kettering.gov.uk/</td></tr>
</table>

Kettering is a market town in the Borough of Kettering, Northamptonshire, England. It is situated about 81 miles (**unknown operator: u'strong'** km) from London. Kettering is mainly situated on the west side of the River Ise, a tributary of the River Nene which meets at Wellingborough. Originally named Cytringan, Kyteringas and Keteiringan in the 10th century, the name Kettering is now taken to mean 'the place (or territory) of Ketter's people (or kinsfolk)'.[2]

As of the last census in 2001, the borough has a population of 81,844[3] whilst the town proper had a population of 51,063[1] and the town is twinned with Lahnstein, Germany and Kettering, Ohio, in the United States. Being part of the Milton Keynes South Midlands (MKSM) study area along with other towns in Northamptonshire, the town is due to get around 6,000 additional homes mainly to the east of the town.[4] The town, like other towns in the area, has a growing commuter population as it is located on the Midland Main Line railway, which has fast InterCity trains directly into London St Pancras International taking around 1 hour. This gives an interchange with Eurostar services to Continental Europe.[5]

Early history

Once believed obscure, the placename Kettering is now taken to mean 'the place (or territory) of Ketter's people (or kinsfolk)'.[2] Spelt variously Cytringan, Kyteringas and Keteiringan in the 10th century, although the origin of the name appears to have baffled place-name scholars in the 1930s, words and place-names ending with 'ing' usually derive from the Anglo-Saxon or Old English word inga or ingas meaning 'the people of the' or 'tribe'.

Before the Romans the Kettering area, like much of Northamptonshire's prehistoric countryside, appears to have remained somewhat intractable with regards to early human occupation, resulting in an apparently sparse population and relatively few finds from the Palaeolithic, Mesolithic and Neolithic periods.[6] About 500 BC the Iron Age was introduced into the area by a continental people in the form of the Hallstatt culture,[7] and over the next century a series of hillforts was constructed, the closest to Kettering being at nearby Irthlingborough.

Roman

Like most of what later became Northamptonshire, from early in the 1st century BC the Kettering area became part of the territory of the Catuvellauni, a Belgic tribe, the Northamptonshire area forming their most northerly possession.[7] The Catuvellauni were in turn conquered by the Romans in 43 AD.

The town traces its origins to an early, unwalled Romano British settlement, the remnants of which lie under the northern part of the modern town. Occupied until the 4th century AD, there is evidence that a substantial amount of iron-smelting took place on the site.[8] Along with the Forest of Dean and the Weald of Kent and Sussex, this area of Northamptonshire "was one of the three great centres of iron-working in Roman Britain".[8] The settlement reached as far as the Weekley and Geddington parishes. However it is felt unlikely that the site was continuously occupied from the Romano British into the Anglo-Saxon era.[9] Pottery kilns have also been unearthed at nearby Barton Seagrave and Boughton.

Saxon

Excavations in the early 20th century either side of Stamford road (A43), near the site of the former Prime Cut factory,(now The Warren public house) revealed an extensive early Saxon burial site, consisting of at least a hundred cremation urns dating to the 5th century AD. This suggests that it may have been among the earliest Anglo-Saxon penetrations into the interior of what later became England. The prefix 'Wic-' of the nearby village of Weekley may also signify Anglo-Saxon activities in the area; Greenall reports that it could be "an indication of foederati, Anglo-Saxon mercenaries brought in to boost the defences of the Empire."[8] This was established imperial policy, which the Romano British continued after Rome withdrew from Britain around 410 AD, with disastrous consequences for the Romano-Britons.

By the 7th century the lands that would eventually become Northamptonshire formed part of the Anglo-Saxon kingdom of Mercia.[10] The Mercians converted to Christianity in 654 AD with the death of the pagan king Penda.[11] From about 889 the Kettering area, along with much of Northamptonshire (and at one point almost all of England except for Athelney marsh in Somerset), was conquered by the Danes and became part of the Danelaw, with the ancient trackway of Watling Street serving as the border, until being recaptured by the English under the Wessex king Edward the Elder, son of Alfred the Great, in 917. Northamptonshire was conquered again in 940, this time by the Vikings of York, who devastated the area, only for the county to be retaken by the English in 942.[12]

It is unlikely however that Kettering itself existed as a village earlier than the 10th century (the county of Northampton itself is not referenced in documents before 1011).[13] Before this time the Kettering area was most likely populated by a thin scattering of family farmsteads.[8] The first historical reference of Kettering is in a charter of 956 AD in which King Edwy granted ten "cassati" of land to Aelfsige the Goldsmith. The boundaries delineated in this charter would have been recognisable to most inhabitants for the last thousand years and can still be walked today. It is possible that Aelfsige the Goldsmith gave Kettering to the monastery of Peterborough, as King Edgar in a charter dated 972 confirmed it to that monastery.

Medieval

At the Domesday survey in 1086, Kettering manor is listed as held by the Abbey of Peterborough, the church owning 10 hides of land. Kettering was valued at £11, with land for 16 ploughs. There were 107 acres (**unknown operator: u'strong'** km^2) of meadow, 3 of woodland, 2 mills, 31 villans with 10 ploughs, and 1 female slave.[14]

The nearby stately home of Boughton House, sometimes described as the 'English Versailles'[15] has for centuries been the seat of the Dukes of Buccleuch, major landowners in Kettering and most of the surrounding villages; along with the Watsons of Rockingham Castle, the two families were joint lords of the manor of Kettering.[16]

Kettering is dominated by the crocketed spire of about 180 feet (**unknown operator: u'strong'** m) of the Parish church of SS Peter and Paul. Little is known of the origins of the church, its first known priest becoming rector in 1219-20. The chancel is in the Early Decorated style of about 1300, the main fabric of the building being mostly Perpendicular, having been rebuilt in the mid 15th century (its tower and spire being remarkably similar to the tower and spire at Oundle). Whether the current building replaced an earlier church on the site is unknown.[7] Two medieval wall paintings, one of two angels with feathered wings, and one of a now faded saint, can still be seen inside the church.[17]

The charter for Kettering's market was granted to the Bishop of Peterborough by Henry III in 1227. In April 1986 the bus station was relocated away from the market area to the Newland Street entrance of the modern Newlands shopping centre, causing a fatal decline in market trade. The ancient market place has since been abandoned. Attempts to revive the market as a small scale street market have met with little welcome from local shops and stores.

17th century

In June 1607 at the nearby village of Newton, the Newton Rebellion broke out,[18] causing a brief uprising known as the Midland Revolt, which involved several nearby villages. Protesting at land enclosures at Newton and Pytchley by local landlords the Treshams, on 8 June a pitched battle took place between Levellers - many from Kettering, Corby and particularly Weldon,[19] - and local gentry and their servants (local militias having refused the call to arms). Approximately 40-50 local men are said to have been killed and the ringleaders hanged, drawn and quartered. The Newton rebellion represents one of the last times that the English peasantry and the gentry were in open conflict. By the 17th century the town was a centre for woollen cloth.

Recent history

The present town grew up in the 19th century with the development of the boot and shoe industry, for which Northamptonshire as a whole became famous. Many large homes in both the Headlands and Rockingham Road were built for factory owners while terraced streets provided accommodation for the workers. The industry has markedly declined since the 1970s,[17] large footwear manufacturers such as Dolcis, Freeman, Hardy and Willis, Frank Wright and Timpsons, having left the town or closed down in the face of stiff overseas competition, while others have outsourced their production to lower-cost countries. Only two smaller footwear businesses remain.[17]

Victorian era Kettering was the centre of the 19th century religious non-conformism and the Christian missionary movement, and this has been preserved in many names. William Carey was born in 1761 at Paulerspury and spent his early life in Kettering before leaving for India as a missionary in 1793. Carey Mission House and Carey Street were named after him. Andrew Fuller helped Carey found the Baptist Missionary Society and he is remembered in the Fuller Church and Fuller Street. In 1803 William Knibb was born in Market Street and became a missionary and emancipator of slaves; he is commemorated by the Knibb Centre and Knibb Street. Toller Chapel and Toller Place are named after two ministers, father and son, who preached in Kettering for a total of 100 years. The chapel was built in 1723 for those who since 1662 had been worshipping in secret.

After several false starts Kettering station was opened in 1857 by the Midland Railway Company, providing a welcome economic stimulus to an ailing local economy, suffering as it was from the loss of wayfaring business since the introduction of railways nationwide. The line was finally linked to London in 1867.[20]

In 1887, John Bartholomew's *Gazetteer of the British Isles* described Kettering as:

> Kettering, market town and parish with railway station, Northamptonshire, 8 miles N. of Wellingborough and 75 miles from London, 2840 ac., pop. 11,095; P.O., T.O.; 3 Banks, 2 newspapers. Market-day, Friday. Kettering is an ancient place, and was called by the Saxons, *"Kateringes"*. It is a fairly prosperous town, with tanning and currying, mfrs. of boots and shoes, stays, brushes, agricultural implements, and some articles of clothing. It has a handsome town hall, a cattle market, a corn exchange, and a grammar school. Many Roman relics have been found in the vicinity.

In 1921 Wicksteed Park, Britain's second oldest theme park, was officially opened on the southern outskirts of the town, and remains popular to this day.

From 1942 to 1945 the town witnessed a large influx of American servicemen (including on several occasions Clark Gable), mainly from the US 8th Air Force at RAF Grafton Underwood, 3.7 miles (**unknown operator: u'strong'** km) away. The base was soon nicknamed 'Grafton Undermud' in reference to the perceived English weather of 'rain, rain and more rain'.[21] The first bombing raid - targeting the marshalling yards at Rouen, northern France - was led by Major Paul W. Tibbets who in 1945 piloted Enola Gay, the B-29 Superfortress that dropped the first atomic bomb on Hiroshima[22] Aircraft from Grafton Underwood dropped the 8th Air Force's first and last bombs of WWII.[23]

Governance

Further information: Northamptonshire

Kettering Borough Council

In local government, Kettering falls within the areas of Northamptonshire County Council and Kettering Borough Council, which incorporates the small, satellite towns of Burton Latimer, Desborough and Rothwell. The borough is split into 17 electoral wards, 10 of these in the town. These comprise of: All Saints, Avondale Grange, Barton, Brambleside, Northfield, Pipers Hill, Ise Lodge, St Michael & Wicksteed and St Peter's.[24]

Kettering Constituency

Kettering is represented in parliament by a constituency of the same name, which is currently (as of May 2010) represented by Conservative[25] MP Philip Hollobone, who gained the marginal constituency from former Labour

MP Phil Sawford in the 2005 general election.

European Parliament

In the European Parliament, Kettering falls within the East Midlands European Parliament constituency and is represented by 5 MEPs.[26]

Historic politics

Politics in Kettering has not always been a sedate affair: in 1835 a horrified Charles Dickens, then a young reporter for the Morning Chronicle, watched aghast as a Tory supporter on horseback, intent (along with others) on taking control of bye-election proceedings, produced a loaded pistol and had to be restrained by his friends from committing murder. The ensuing riot between Tory and Whig supporters led Dickens in his article to form various opinions of Kettering and its voters, none of them complimentary.[27]

Local economy and amenities

Kettering's economy was built on the boot and shoe industry. With the arrival of railways in the 19th century, industries such as engineering and clothing grew up. The clothing manufacturer Aquascutum built its first factory here in 1909. Now Kettering's economy is based on service and distribution industries due to its central location and transport links. Recent times have seen, directly linked with the economic downturn many of the shops have closed down, predominantly localised, smaller shops leaving the town centre somewhat diminished.

A derelict shoe factory

Kettering's unemployment rate is amongst the lowest in the UK and has over 80% of its adults in full time employment.[28] It is home to a wide range of companies including Weetabix, Pegasus Software, RCI Europe, Timsons Ltd and Morrisons Distribution as well as Wicksteed Park, the United Kingdom's oldest theme park, which now plays host to one and a quarter million visitors every season.

Kettering is the home of Kettering General Hospital,[29] which provides Acute and Accident & Emergency department services for north Northamptonshire including Corby and Wellingborough. With its new £20 million campus,[30] 16,000 students and 800 staff, Tresham College of Further and Higher Education is a significant employer in the region.

Kettering Business Park, a recent and current commercial property development undertaken by Buccleuch Property is situated on the A43/A6003, on the north side of Kettering.[31] Many office buildings are being built as part of the project as well as a leisure sector with a new hotel. Many large distribution warehouses have been constructed in the area, creating thousands of jobs for the local economy. Kettering's Heritage Quarter houses the Manor House Museum and the Alfred East Gallery.[32] The magnificent Boughton House, Queen Eleanor cross and the 1597 Triangular Lodge are local landmarks within the borough. Sir Thomas Tresham was a devout Catholic who was imprisoned for his beliefs. When he was released he built Triangular Lodge to defy his prosecutors and secretly declare his faith. The construction's 'three of everything' - sides, floors, windows and gables - represent the Holy Trinity.

The British sitcom "Peep Show" has various scenes located in Kettering owing to the head office of JLB, the company which employs lead character Mark Corrigan, being located there. However, the scenes are not filmed in Kettering, and places named in the show such as the nightclub *Lap Land Kettering*, and the hotel *Park Kettering* are fictitious.

Sport

Football

Kettering was home to Kettering Town F.C. The current Chairman Imraan Ladak[33] installed former Tamworth manager Mark Cooper as the new Kettering Town supremo for the 2007-08 season. After a record breaking start to the season (7 consecutive wins), the club held pole position virtually all season, winning the Blue Square North with 5 games in hand. The Poppies broke their win record of 28 games, now 30 and registered a record points tally for the Division of 96 points. Kettering Town play in the Blue Square Premier. Mark Cooper joined Peterborough United Manager in November 2009 with Goalkeeper Lee Harper taking over as he was the most experienced player. However, as of 2011, the club moved out of the borough to Nene Park, Irthlingborough - the former ground of defunct rivals Rushden & Diamonds. The move was necessitated by the approaching end of the lease at their old ground at Rockingham Road, Kettering. Mark Cooper returned as caretaker manager until the end of the current (2011–2012) season on 4 January 2012. He replaced former manager Mark Stimson.

Rugby

Kettering is home to Kettering Rugby Football Club (KRFC), located in Waverley Road on the eastern side of the town.[34] The earliest available records indicate that the playing of Rugby Football in Kettering was initiated by the Rector of Barton Seagrave village in 1871. After a period of playing under Uppingham Public School Rules the club formally adopted RFU rules in 1875 and quickly became a significant participant in both the local community and the fast-developing Rugby scene in the East Midlands. In the early days games were played on a number of sites including farmers' fields and council-owned grounds. It was during this period, prior to adopting a home of their own, that the club developed its high profile in the town. Social occasions and players "meetings" were held traditionally at the Royal Hotel, later moving to the George, with more formal occasions such as the Annual Ball becoming the highlight of the local function calendar. KRFC currently plays in the Midlands 1st Division.

Transport

Roads

The A14 skirts the west and south of the town, links the town with the A45 dual carriageway, M1 and M6 motorways. The A43 links Kettering with the county town of Northampton and the A509 (Kettering/Wellingborough Road) links Kettering with Wellingborough.

Buses

In April 1986 the bus station was relocated away from the market area to the Newland Street entrance of the modern Newlands shopping centre, causing a fatal decline in market trade. Although buses were re-allocated there in April 1987 before closing again in September 1989 and building a smaller version of a Bus Station on which it closed also in May 1999, buses since then have just served The Library and Newlands Shopping Centre. From May 2010 all buses now serve the new horsemarket bus interchange as they don't serve the library any more this is due to public realem improvements. New bus stops have been installed around rail station and headlands.

The Town is also served by a local bus network under the brand name *Connect Kettering* with routes A,B,C,D, E and F linking the town centre with local suburbs and Burton Latimer.[35] Leaving every 30 minutes, the X4 service links the town with Milton Keynes, Northampton, Wellingborough, Corby, Oundle and Peterborough.[36]

Rail

Rail services operated by East Midlands Trains depart every 30 minutes from Kettering to St Pancras International railway station, with an average journey time of 59 minutes.[37] St Pancras also provides an interchange with the Eurostar service to France and Belgium. Kettering is linked to Corby, Leicester, Nottingham, Derby and Sheffield to the north and Wellingborough, Bedford, Luton to the south.[38] Because of good rail links, a large and growing commuter population takes advantage of Kettering's position on the Midland Main Line railway.

Kettering station platform

Airports

Five large UK airports are within 2 hours' drive of the town, these being London Heathrow, London Luton, East Midlands, Birmingham International and London Stansted. London Luton can be reached directly by train while East Midlands and London Stansted can be reached by one change at Leicester. Sywell Aerodrome, located 6 miles (**unknown operator: u'strong'** km) south-west of Kettering, caters for private flying, flight training and corporate flights.

Notable residents

- Frank Bellamy - Illustrator of comics
- William Carey - Missionary
- Richard Coles - Member of 80s band The Communards
- Sir Alfred East - Painter
- John Alfred Gotch - Architect and Architectural historian
- Thomas Cooper Gotch - Illustrator
- Sienna Guillory - Actress
- William Knibb - Missionary
- John Profumo - Former British politician, best known for the Profumo Scandal
- Edward Sismore - Air Commodore
- Charles Wicksteed - Created Wicksteed Amusement Park
- Andrew Kooner - Canadian boxer, born in Kettering
- Russ Russell - Professional music engineer
- Hugh Dennis - English comedian, born in Kettering
- Jane Clarke - Children's Author, born in Kettering
- Reginald Wooster - Cricketer, born and died in Kettering
- Ernest Wright - Cricketer, born and died in Kettering

Future growth

In mid-2003 the population of the Borough of Kettering was estimated at 86,000, with 51,063 residing in the town proper. Kettering along with the neighbouroughing town of Corby is located in North Northamptonshire, the fastest growing place in England and Wales,[39] with the East Kettering development area covers 300 hectares and extends from the A43 in the north to the A14 in the south.[40] In March 2007, a project was revealed to refurbish and bring new leisure and shopping to the town centre, including water features, public art, sculptures, street furniture, trees, plants and an innovative pavement lighting scheme.[41]

Geography

Climate

Kettering experiences an oceanic climate (Köppen climate classification) which is similar to most of the British Isles.

Compass

Kettering's nearest towns are Desborough, Burton Latimer and Rothwell with the larger towns of Corby and Wellingborough a bit further away.

Town twinning

Kettering is twinned with:

- Kettering, Ohio, USA
- Lahnstein, Germany

See also

- Kettering Ironstone Railway
- Kettering Grammar School

References

[1] http://www.statistics.gov.uk/statbase/Expodata/Spreadsheets/D8271.csv

[2] R.L. Greenall: A History of Kettering, Phillimore & Co. Ltd, 2003, ISBN 1-86077-254-4. p.7.

[3] UK Statistics: Kettering (http://www.statistics.gov.uk/census2001/pyramids/pages/34ue.asp) Retrieved 3 May 2010

[4] Live in Kettering:3D Map (http://www.liveinkettering.co.uk/town-centre-vision/3D-map) Retrieved 1 May 2010

[5] East Midlands Trains: Interchange with Eurostar (http://www.eastmidlandstrains.co.uk/EMTrains/YourDestinations/Eurostar/) Retrieved 1 May 2010

[6] R.L. Greenall: A History of Northamptonshire, Phillimore & Co. Ltd, 1979, ISBN 1-86077-147-5. p.19.

[7] R.L. Greenall: A History of Northamptonshire, Phillimore & Co. Ltd, 1979, ISBN 1-86077-147-5. p.20.

[8] R.L. Greenall, A History of Kettering, Phillimore & Co. Ltd, 2003, ISBN 1-86077-254-4. P. 9.

[9] R.L. Greenall, A History of Kettering, Phillimore & Co. Ltd, 2003, ISBN 1-86077-254-4. P. 10.

[10] R.L. Greenall: A History of Northamptonshire, Phillimore & Co. Ltd, 1979, ISBN 1-86077-147-5. p.26.

[11] R.L. Greenall: A History of Northamptonshire, Phillimore & Co. Ltd, 1979, ISBN 1-86077-147-5. p.29.

[12] Michael Wood: The Domesday Quest, BBC Books, 1986 ISBN 0-563-52274-7. p. 90.

[13] R.L. Greenall: A History of Northamptonshire, Phillimore & Co. Ltd, 1979, ISBN 1-86077-147-5. p.34.

[14] Domeday Book: A Complete Translation, Penguin Books, 1992, ISBN 0-14-100523-8. p. 596.

[15] Michael McNay: Hidden Treasures of England, Random House Books, 2009, ISBN 1-905211-83-X. p.271.

[16] R.L. Greenall: A History of Kettering, Phillimore & Co. Ltd, 2003, ISBN 1-86077-254-4. p.4.

[17] R.L. Greenall, A History of Kettering, Phillimore & Co. Ltd, 2003. p.21.

[18] R.L. Greenall: A History of Northamptonshire, Phillimore & Co. Ltd, 1979, ISBN 1-86077-147-5. p.41-42.

[19] R.L. Greenall: A History of Northamptonshire, Phillimore & Co. Ltd, 1979, ISBN 1-86077-147-5. p.42.

[20] R.L. Greenall, A History of Kettering, Phillimore & Co. Ltd, 2003. p.116.

[21] John N. Smith, Airfield Focus 44: Grafton Underwood, GMS Enterprises, 2001, ISBN 1-870384-84-9. p. 4.

[22] John N. Smith, Airfield Focus 44: Grafton Underwood, GMS Enterprises, 2001, ISBN 1-870384-84-9. p. 3.

[23] John N. Smith, Airfield Focus 44: Grafton Underwood, GMS Enterprises, 2001, ISBN 1-870384-84-9. p. 33.

[24] Kettering Borough Council: Wards 2009 (http://www.kettering.gov.uk/downloads/Wards_2009.pdf) Retrieved 3 May 2010

[25] Kettering Conservatives (http://www.kettering-conservatives.org/) Retrieved 3 May 2010

[26] East Midlands MEPs (http://www.europarl.org.uk/section/your-meps/findmep?filter0=East+Midlands&filter1=**ALL**& filter2=**ALL**&filter3=**ALL**&filter4=**ALL**) Retrieved 3 May 2010

[27] R.L. Greenall, A History of Kettering, Phillimore & Co. Ltd, 2003, ISBN 1-86077-254-4. P. 106.

[28] Government Office of the East Midlands (http://www.go-em.gov.uk/geographical.php?LA=34UE&x=0&county=northants&y=1)

[29] "Kettering General Hospital: Getting here" (http://www.kgh.nhs.uk/getting-here). Kettering General Hospital, NHS Foundation Trust. . Retrieved 30 July 2010.

[30] "Tresham College of Further and Higher Education: Kettering" (http://www.tresham.ac.uk/about/our_campuses/kettering). Tresham College of Further and Higher Education. . Retrieved 30 July 2010.

[31] "Kettering Business Park - Overview" (http://www.ketteringbusinesspark.com/overview.php). Kettering Business Park. . Retrieved 30 July 2010.

[32] "A Walk Around Kettering's Heritage Quarter" (http://www.kettering.gov.uk/downloads/walk_heritage_quarter.pdf). . Retrieved May 25, 2009.

[33] "Kettering Town FC: Club Information" (http://www.ketteringtownfc.co.uk/club_info.php). Kettering Town FC. . Retrieved 30 July 2010.

[34] "Kettering Rugby Football Club: How to find us" (http://www.ketteringrugbyclub.co.uk/about/howtofindus.html). Kettering Rugby Football Club. . Retrieved 30 July 2010.

[35] "Stagecoach in Northants: Connect Kettering Network us" (http://www.stagecoachbus.com/uploads/ketteringmapdownloadlo.pdf). Stagecoach Group plc. . Retrieved 30 July 2010.

[36] "Stagecoach in Northants: X4" (http://www.stagecoachbus.com/northamptonshire-X4.aspx). Stagecoach Group plc. . Retrieved 30 July 2010.

[37] "East Midlands Trains: Midland Main Line (London services) Timetable" (http://eastmidlandstrains.co.uk/emtrains/images/pdfs/TT1May10PagesandTablesv3DH(2).pdf). East Midlands Trains - Stagecoach Group plc. . Retrieved 30 July 2010.

[38] "East Midlands Trains: Network Map" (http://www.eastmidlandstrains.co.uk/EMTrains/YourDestinations/OurNetworkAndStations.htm). East Midlands Trains - Stagecoach Group plc. . Retrieved 30 July 2010.

[39] "Corby - fastest growing place in England and Wales" (http://www.corby.gov.uk/CouncilAndDemocracy/CouncilNews/Pages/CorbyâfastestgrowingplaceinEnglandandWales.aspx). Corby Borough Council. . Retrieved 30 July 2010. "The statistics highlight North Northamptonshire's emergence as the biggest single growth area outside London"

[40] "East Kettering: Land use map" (http://www.eastkettering.com/Land_Use_Plan_2008.html). . Retrieved 30 July 2010.

[41] "A new dawn" by Monique Cleaver in *Northants Evening Telegraph*, 21 March 2007 (http://www.ketteringtoday.co.uk/ViewArticle.aspx?ArticleID=2135857&SectionID=5388)

External links

- Kettering local free business directory, local news, events and offers (http://www.ketteringlocal.com/)
- Kettering Weather Station - Supplying local weather for local people (http://ketteringweathercentre.co.uk/)
- BBC Northamptonshire (http://www.bbc.co.uk/northamptonshire/)
- Kettering Evening Telegraph (http://www.ketteringtoday.co.uk/)
- Kettering Borough Council (http://www.kettering.gov.uk/)
- Kettering Town F.C. (http://ketteringtownfc.co.uk/)
- Old Cytringanians (http://www.cytringanians.co.uk/)
- History Notes - The Kettering Grammar School Satellite Tracking Group (http://www.zarya.info/Kettering/Kettering.php)

Market_Harborough

<table>
<tr><th colspan="2" style="text-align:center">Market Harborough</th></tr>
<tr><td colspan="2" style="text-align:center">
The Old Grammar School</td></tr>
<tr><td colspan="2">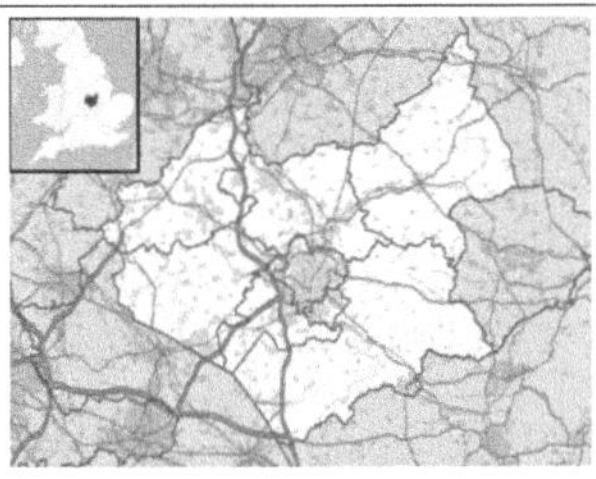
Market Harborough
Market Harborough shown within Leicestershire</td></tr>
<tr><td>Population</td><td>Expression error: "20,785" must be numericTemplate:Infobox UK place/trap[1]</td></tr>
<tr><td>OS grid reference</td><td>SP7387</td></tr>
<tr><td>- London</td><td>88 miles (unknown operator: u'strong' km)</td></tr>
<tr><td>District</td><td>Harborough</td></tr>
<tr><td>Shire county</td><td>Leicestershire</td></tr>
<tr><td>Region</td><td>East Midlands</td></tr>
<tr><td>Country</td><td>England</td></tr>
<tr><td>Sovereign state</td><td>United Kingdom</td></tr>
<tr><td>Post town</td><td>MARKET HARBOROUGH</td></tr>
<tr><td>Postcode district</td><td>LE16</td></tr>
<tr><td>Dialling code</td><td>01858</td></tr>
<tr><td>Police</td><td>Leicestershire</td></tr>
<tr><td>Fire</td><td>Leicestershire</td></tr>
<tr><td>Ambulance</td><td>East Midlands</td></tr>
<tr><td>EU Parliament</td><td>East Midlands</td></tr>
<tr><td>UK Parliament</td><td>Harborough</td></tr>
</table>

Market Harborough is a market town within the Harborough district of Leicestershire, England. It has a population of 20,785[1] and is the administrative headquarters of Harborough District Council.[2] It sits on the Northamptonshire-Leicestershire border. The town was formerly at a crossroads for both road and rail; however the A6 now bypasses the town to the east and the A14 which carries east-west traffic is 6 miles (**unknown operator: u'strong'** km) to the south. The town is served by East Midlands Trains with direct services to Leicester, Nottingham, Derby and St Pancras International. Rail services to Rugby and Peterborough ended in 1966.

St Dionysius Church with the Old Grammar School (right)

Market Harborough is located in an area which was formerly a part of the Rockingham Forest, a royal hunting forest used by the medieval monarchs starting with William I. Rockingham Road takes its name from the forest. The forest's original boundaries stretched from Market Harborough through to Stamford and swallowed up Corby, Kettering, Desborough, Rothwell, Thrapston and Oundle.

The centre of the town is dominated by the steeple of St. Dionysius Parish Church which rises directly from the street, as there is no church yard. It was constructed in grey stone in 1300 with the church itself a later building of about 1470. Next to the church stands the Old

St Mary's Place, Market Harborough

Grammar School, a small timber building dating from 1614. The ground floor is open, creating a covered market area and there is a single room on the first floor. It has become a symbol of the town. The nearby square is largely pedestrianised and surrounded by buildings of varying styles. The upper end of the High Street is wide and contains mostly unspoiled Georgian buildings.

Market Harborough has two villages within its confines: Great Bowden lies over a hill about a mile from the town centre; Little Bowden is less than half a mile from the town centre. The three centres have largely coalesced through ribbon development and infill, although Great Bowden continues to retain a strong village identity.

History

1200-1799

Domesday Book records Bowden as a Royal Manor organised in three manors. The population lived in three villages, Great Bowden, Arden and Little Bowden. The Manor of Harborough is first mentioned in 1199 and 1227 when it was called "Haverberg". It is likely that Harborough was formed out of the Royal Manor with the intention of making it a place for tradesmen and a market when a new highway between Oxendon and Kibworth was established to help link Northampton and Leicester. A chapel dedicated to St Dionysius was built on the route, whilst St Mary in Arden retained Parish Church status.

The name of Harborough is likely to derive from the Anglo-Saxon "haefera-beorg" or oat hill. A market was established by 1204 and has been held on a Tuesday ever since 1221. The trades people of Harborough had large tofts or farm yards at the rear of their property where goods were made and stored. Many of these yards remain but have been subdivided down their length over the years to give frontage to the High Street

The Steeple of Harborough Church was started in 1300 and completed in 1320. It is a broache spire, which rests on the walls of the tower, and are earlier than recessed spires which rise from behind a square tower as at Great Bowden. By 1382 the village of Arden had been abandoned, although the Church remained in use for some years. In 1470 the main part of Harborough Church was completed. An open stream ran down the High Street. The Town Estate was created and managed by a body of Feoffees elected by the townspeople, to help manage among other things the Open Fields surrounding the town, the proceeds from which were used for a variety of purposes. From 1570 the Town Estate owned several properties within the town.

Harborough figured nationally in the English Civil War in June 1645, when it became the headquarters of the King's Army. In Harborough, the King decided to confront Parliamentary forces who were camped near Naseby but the Battle of Naseby proved a decisive victory for Parliament led by Oliver Cromwell. Harborough Chapel became a temporary prison for the captured forces. Cromwell wrote a letter from "Haverbrowe, June 14, 1645" to the Speaker of the House of Commons, William Lenthall, announcing the victory.

An Independent Church was established in the Harborough area following the 1662 Act of Uniformity and a Meeting House was built in Bowden Lane in 1694.

During the 18th century the timber mud and thatch buildings of the town were largely replaced with brick buildings. After roads were turnpiked and regularly repaired (making wheeled traffic easier all year round) Harborough became a staging point for coach travel on the road to London from the North West and the Midlands. In 1776 the Open Fields of Great Bowden were allotted to individual owners and fenced with hedges planted, followed by those of Little Bowden in 1780.[3]

1800-1899

In the 19th century, the increasing level of heavy goods traffic on the turnpike roads led to complaints. A plan for a canal from Leicester to join the London to Birmingham canal was mooted but it eventually bypassed the town and a branch canal was cut from Foxton to Harborough with wharves at Gallow Hill, and Great Bowden. Harborough wharf, to the north of the town, became a distribution centre for coal and corn. A gas company was formed in 1833 to make and distribute gas. John Clarke and Sons of London built a factory for spinning worsted and later making carpets. Other industries developed were a brickworks, brewery, wheelwright/coachworks and the British Glues and Chemicals works by the Canal at Gallow Hill. In the 1830s a union of parishes around Market Harborough was formed to look after the poor and a workhouse was built in 1836 on the site of St Luke's Hospital. In 1841 Thomas Cook who was a wood turner and cabinet maker in the town organised the first group travel by rail from Leicester to Loughborough and went on to found the travel agency bearing his name.

Market Harborough became a centre for fox hunting with hounds during the 19th century when Mr Tailby of Skeffington Hall established a hunt in South East Leicestershire in 1856. The country between Billesdon and Harborough was considered severe, involving jumping the specially designed ox fences. His hunting diary is recognised as an important document in the history of hunting. The Hunt was renamed the Fernie after a subsequent Master.

The Grand National Hunt Steeple Chase was held to the south west of the town in 1860, 1861 and 1863. This race and the meeting eventually developed into the Cheltenham Festival and the organisers were part of the founding of organised steeplechasing through the Grand National Hunt Committee [4]

The building of the Leicester–Rugby railway in 1840 had a catastrophic effect on the coaching traffic through the town. A railway did not serve the town until 1850 with a link to Rugby but this was quickly followed by links to Leicester and London in 1857 and to Northampton in 1859.

In 1850 William Symington [5], a grocer in the town established a factory to make pea-flour. His brother James developed a haberdashery and stay making business and in 1876 his sons acquired the old carpet factory to make corsets. They expanded it by three additional floors in 1881 and then built a new factory opposite Church Square in 1884 which still remains today as the Council offices, library and museum. In the 1890s Harborough Rubber Company, Looms Wooden Heels and Caxton Die Casting works were established. A Tannery was built on land adjoining the Commons.

There was a rapid expansion in the town's population from 4,400 in 1861 to 7,700 in 1901. This had been at the expense of living conditions with severe overcrowding in the old town.. Rows of cottages had been built in the yards of older houses with shared access to water and waste disposal. The Public Health Act of 1875 required local authorities to implement building regulations, or bye-laws, which insisted that each house should be self-contained, with its own sanitation and water. In 1883 a new system of sewers were laid and piped water supplied from wells at Husbands Bosworth. Additional residential areas were developed – the New Harborough estate off Coventry Road and the Northampton Road estate between Nithsdale Avenue and Caxton St.

1900-1999

In 1888 Little Bowden parish was transferred from Northamptonshire to Leicestershire and following the Local Government Act of 1894, an Urban District Council was formed for Market Harborough, covering the town and the parishes of Little and Great Bowden. Various schemes were implemented to improve the town. It acquired the gas company and built a public baths. It acquired land for the construction of Abbey Street in 1901 which removed the a multi occupied yard of the Coach and Horses Inn and enabled the building of a fire station on the new Street in 1903. In the same year a new livestock market was opened between Springfield Street and the river on 12 acres (**unknown operator: u'strong'** m^2) of land, enabling the cattle and sheep markets to be cleared from the streets. In 1905 the council bought land at Great Bowden and Little Bowden for recreation grounds.

In 1919 there were still around 150 dwellings identified as unfit for human habitation mostly in the yards and courts of Harborough and there was an identified need for 300 new houses. Land to the north of the town was selected and a scheme for 98 homes for rent developed as the Bowden Fields Estate. Following the introduction of mortgage subsidy, over 100 private homes were built and a further development of 72 rented homes took place. By 1928 about 400 houses had been built since 1918, 164 by the Council. A major improvement took place from 1930 with the acquisition of land between Northampton Road and Farndon Road. This enabled the construction of Welland Park Road (which enabled east west traffic to bypass the town centre), provision of 100 homes for rent along Welland Park Road and 52 in Walcot Road to rehouse occupants of the old yard houses, plots for private housing, the layout of Welland Park and the construction of Welland Park School.

A covered Market Hall was opened at the western end of the Cattlemarket in 1938, replacing the market stalls on the Square on Tuesdays and Saturdays.

The postwar period saw another shortage of housing and some 600 people on the waiting list for social housing. The council developed a 100 dwelling extension to the Bowden Fields Estate by 1949 and acquired 140 acres (**unknown operator: u'strong'** km^2) of land to the south west of the town to deal with the problem. A new Southern Estate was planned to accommodate 700 dwellings, shopping centre, school and recreation ground. The Council laid initial access roads named after personalities of the Battle of Naseby since these fields were crossed by both armies on 14 June 1645. A plaque now records the events and was unveiled by Mrs H.B. Lenthall on 1 February 1951 to mark the opening of the Estate development. Around 150 dwellings were built for rent with the remaining plots available for private building. The final phase of development occurred in the 1980s.

In 1950 the canal basin was the venue for a week long National Festival of Boats, the first such Festival organised by the Inland Waterways Association and marking the beginning of the revival of the canal network for leisure use. The old Brewery site was acquired for a bus station in 1951 and in 1958 a main car park was opened at the Commons and further car parks established in the 1960s to deal with the increasing demand. Proposals for development of an

Industrial Estate at Riverside and Rockingham Road were approved in 1962 and the area developed during the 1960s.

Following serious flooding in the town centre on 2 July 1958, a flood relief scheme was begun and the river bed was straightened and deepened.

In 1968 the centre of Market Harborough was declared a conservation area. Major developments included the development of headquarters for Golden Wonder crisp makers, and the demolition of the old Symington factory in Adam and Eve Street for redevelopment as Eden Court shops and flats.

During the 1970s, draft proposals were made for an inner relief road to avoid traffic congestion in the town centre. However, it was rejected in favour of a bypass outside the town.[6]

In 1980 the Symingtons Factory at Church Square was redeveloped as the District Council offices, Library and Museum. Plans for an A6 by-pass were approved by the Department for Transport during the 1980s and the 5 miles (**unknown operator: u'strong'** km) road costing £9.5m was opened in June 1992. In addition, proposals were made for a new east-west link road (A14) between the A1 and M1 and a route was identified 10 miles (**unknown operator: u'strong'** km) south. It was opened in summer 1991. The opening of these roads has reduced considerably the volume of heavy goods vehicles passing through the town centre.

Associated improvements to the town centre took place as part of a "By-pass Demonstration Project" completed in 1994. This involved comprehensive re-paving and new street furniture to make the centre more pedestrian friendly whilst through-traffic with a 20 mph (**unknown operator: u'strong'** km/h) speed limit.

In 1993 the former cattle market, bus station, indoor market and several properties next to the old post office and the Peacock Hotel were re-developed to form a new pedestrianised shopping centre called St Mary's Place. This included a Sainsburys supermarket.

2000 onwards

A footpath and cycleway alongside the canal to Foxton became part of the National Cycle Network Route 6. The path continues south following the Brampton Valley Way a long and narrow recreation area on the route of the former railway line to Northampton.

Canal Basin

The Canal Basin was restored as a boating centre called Union Wharf. This consists of workshops, restaurant, studios and apartments. There are residential moorings and canal boats can be hired.

A cycle and footway along the river through the town was created called the Millennium Mile and links Welland Park with the railway station. In 2007 Welland Park was awarded Green Flag Award status and in 2008 a large new children's play area was opened.

Geography

Market Harborough is in a rural part of south Leicestershire, on the River Welland and close to the Northamptonshire border. The town is about 15 miles (**unknown operator: u'strong'** km) south of Leicester via the A6, 17 miles (**unknown operator: u'strong'** km) north of Northampton via the A508 and 10 miles (**unknown operator: u'strong'** km) north west of Kettering. The town is near the A14 road running from the M1/M6 motorway Catthorpe Interchange to Felixstowe. The M1 is about 11 miles (**unknown operator: u'strong'** km) west via the A4304 road.

The Midland Main Line railway connects to London St Pancras International. A branch of the Grand Union Canal terminates in the north part of the town and connects to the main canal near Foxton and the Foxton Locks. (see below)

Governance

The Parliamentary Constituency is Harborough which includes the town, surrounding rural areas as well as the urban areas of Oadby, Wigston and South Wigston, all southern suburbs of the city of Leicester to the north. The MP as of the late first decade of the 21st century is Edward Garnier.

The District Council is Harborough District Council,[2] with its offices in the town centre being the former Symingtons Corset Factory. The town itself is however an unparished area, with no town council of its own - the third least populated town of this sort.

The town is in the southern area of Leicestershire County Council close to the border with Northamptonshire.

Landmarks

The Old Grammar School

One of the town's most notable features is an unusual former grammar school located in the town centre which stands on wooden stilts. The school room had to be built upon posts to allow the butter market to be held on the ground floor. The School was founded in 1607 and built in 1614, through the generosity of Robert Smyth, a poor native of the town who became Comptroller of the Lord Mayor's Court of the City of London and member of the Merchant Taylors' Company.

The subjects taught were Latin, Greek and Hebrew, and many boys were sent to Oxford and Cambridge Universities. The most distinguished of these was John Moore, who became Bishop of Norwich in 1691, and Bishop of Ely in 1707 and also William Henry Bragg, Nobel Prize winner. This is commemorated by a plaque inside the old schoolroom.

The grammar school has since moved sites and is now the Robert Smyth School for 14- to 18-year-olds. The school badge is the arms of the City of London. The school is divided into houses one of which is named "Bragg".

Running around the building are five portions of scripture from the Bible. They are:

> The Lord build the house, they labour in vain that build it: except the Lord keep the city, the watchman waketh but in vain

> —Psalm 127:1

> He the Lord seeth not as man seeth; for man looketh on the outward appearance, but the Lord looketh on the heart

> —1 Samuel 16:7b

> I was glad when they said unto me, Let us go into the house of the Lord

> —Psalm 122.1

> Seek ye first the kingdom of God, and his righteousness; and all these things shall be added unto you

> —Matthew 6.33

> For by grace are ye saved through faith; and that not of yourselves: it is the gift of God:not of works, lest any man should boast

> —Ephesians 2:8-9[7]

St Dionysius Church

- Description and Image [8]

The Church of England parish church is dedicated to Saint Dionysius, and is of a broach spire construction. It dates back to the 14th century but has been added to since.

The Old Town Hall

- Description and Image [9]

The Old Fire Station, Abbey Street

- Description and Image [10]

42 High Street

- Description and Image [11]

Harborough Museum

The Harborough Museum is in part of what was once Symington's Corset Factory, and shares the building with the council offices and library. The museum opened in 1983 and collect and display objects of local interest including local Roman era archaeological finds. It is open to the public most days and admission is free.

St Mary in Arden Chapel

- Description and image [12]

St Mary's Place and the Settling Rooms

In 1993 the former cattle market, bus station, indoor market and several properties adjacent to the old post office and the Peacock Hotel were re-developed to form a new pedestrianised shopping centre called St Mary's Place. It has 26 retail units, and a new indoor market hall, including a Sainsbury's supermarket. The original 'Settling Rooms' from the cattle market have been listed for preservation in the centre of the car park Description and Image [13]. The site straddles the River Welland, a pedestrian suspension bridge, and two other footbridges. There is a Post Office combined with a newsagents and shops. The Peacock is at the main entrance.

Former Flour Mills, St Mary's Road

- Description and Image [14]

Places nearby

Foxton Locks

Three miles north west of the town is Foxton Locks - ten canal locks consisting of two "staircases" each of five locks, on the Leicester line of the Grand Union Canal. It is named after the nearby village of Foxton where there is one of a very few remaining road swing bridges over the canal.

Gartree Prison

HM Prison Gartree is west of the town near Foxton and the site of a prisoner escape by helicopter in 1987. The prison caters for prisoners on life sentences.

Economy

There are 4,750[15] VAT or PAYE registered businesses in the Harborough district. Compared to the United Kingdom the Harborough district has a greater proportion of smaller organisations with fewer than 10 employees; 87.16% vs. 82.8% in the UK overall.

CDS Global have their UK office at Tower House on Sovereign Park, off the A508 - *Northampton Road* towards the leisure centre. They are a data management company, mostly dealing with magazine subscriptions - known as customer relationship management (CRM). The company is owned by the Hearst Corporation who publish magazines such as *Cosmopolitan* and *Esquire*. Hearst Magazines UK have their UK shop address at the Market Harborough office. The worldwide head office is in Des Moines, Iowa.

Haddon-Oldham were based in the town, which produced lead-acid batteries (made in Arras in north France), who are now based in Chippenham in Wiltshire.

Golden Wonder was based at *Edinburgh House* from 1970 until 2006 when it went into administration under Kroll. Not that much further north was its main competitor, responsible for its journey into the red, who now enjoy market domination. Before 1970 it was based in Corby. The former headquarters is about to become a Travelodge.

Abtec Network Systems Ltd has its head office in Market Harborough in the Compass Point Business Park. It's one of the oldest IT firms in the Leicestershire area, established in January 1991.

The Leicestershire, Northamptonshire and Rutland branch of the NFU was in the town from 1975. York Trailer Company were based in the town (and Corby).

Sport and leisure

Football: Market Harborough has two teams: Harborough Town and Borough Alliance. Both cater for a variety of ages. Harborough Town FC has three senior teams, including a ladies' team. The Northampton Road clubhouse has received Football foundation, council and Bowden's Charity grants and awards, as well as sponsorship money, for improvements. Borough Alliance FC was founded in 2003 with a home ground at Symington's recreation ground.

Squash and Racketball Club. Located in Fairfield Road, it has 3 courts and a bar area. Two mens teams in Leicestershire Leagues, 3 teams in the Northants League and a Ladies Team in Leicestershire. Junior coaching with a junior team which shares its facilities with MHCC.(www.harboroughsquash.co.uk)

Cricket: Market Harborough Cricket Club has two cricket teams; Market Harborough CC and Harborough South CC. The former plays in the Leicestershire Premier Cricket League

Rugby: Market Harborough Rugby Club is near the leisure centre and until recently known as Kibworth Rugby Union Club.

Golf: The Market Harborough Golf Club sits to the south of the town itself; much of the golf course actually crosses over into Northamptonshire and is only about a mile from the Northants village of Great Oxendon. It is an 18-hole course and was set up in 1898.

Leisure: A leisure centre has a swimming pool, gym, sauna, steam room and cafe and is open to members and non-members. There are two skateparks: one in Little Bowden and one in the grounds of the youth centre on Farndon Road. The town once had a cinema known as The Ritz, from 1939 till 1978. Now closed, there is a campaign to have it re-opened (www.harboroughcinema.com). The town has a nightclub called Enigma [16].

Transport

Market Harborough station is on the Midland Main Line and operated by East Midlands Trains. London St Pancras International is 70 minutes south. Northbound trains operate to Leicester (15 minutes), Nottingham, Sheffield, Leeds and York. Leicester connections east and west. From November 2007 St Pancras has Eurostar services to the continent.

Media

HFM on 102.3FM was formed in November 1994 to provide a local FM station for Market Harborough and South Leicestershire, as it was felt the established local independent and BBC stations did not cater for the area.

OFCOM awarded the station a full time licence on 15 July 2005 for 24 hours, 365 days a year operation. It is run largely by volunteers but has some freelance presenters.

The station launched full time on Saturday 10 February 2007, with a live broadcast from The Square in Market Harborough's town centre. The first official voice on the station was that of Chris Jones, Programme Controller, and the first record played was "Are You Ready For Love" by Elton John.

HFM's programming is music-based but also focuses on community news and events. There is a daily Community Update as well as local news on the hour 8:00 am – 5:00 pm on weekdays, with updates at the weekends. Outside of these times, the station takes IRN (Independent Radio News)

The *Harborough Mail* is the local newspaper published each Thursday.

Market Harborough Magazine [17] is a is a glossy monthly publication covering Market Harborough and surrounding area

References

[1] 2001

[2] "Harborough District Council website" (http://www.harborough.gov.uk/dotGov/default.jsp/). . Retrieved 2008-12-25.

[3] Bowden to Harborough J.C.Davies 1964

[4] Stevens, Peter, History of the National Hunt Chase 1860-2010, pp 3-20. ISBN 978-0-9567250-0-4.

[5] http://www.symingtons.com/about-us

[6] "Town Affairs - the making of modern Market Harborough" J C Davies 1974

[7] Churches Together in Harborough brochure 2006

[8] http://www.imagesofengland.org.uk/details/default.aspx?id=189621

[9] http://www.imagesofengland.org.uk/details/default.aspx?id=189692

[10] http://www.imagesofengland.org.uk/details/default.aspx?id=402533

[11] http://www.imagesofengland.org.uk/details/default.aspx?id=189681

[12] http://www.imagesofengland.org.uk/details/default.aspx?id=189642

[13] http://www.imagesofengland.org.uk/details/default.aspx?id=402545

[14] http://www.imagesofengland.org.uk/details/default.aspx?id=402532

[15] UK Business: Activity, Size and Location - 2011; Office of National Statistics 2011; http://www.ons.gov.uk/ons/rel/bus-register/uk-business/2011/index.html

[16] http://www.club-enigma.co.uk/

[17] http://www.marketharboroughmagazine.co.uk

External links

- Market Harborough Community website (http://www.marketharborough.com/)
- Harborough District Council (http://www.harborough.gov.uk/)
- Photographs of Market Harborough and area (http://www.geograph.org.uk/search.php?i=2750925/)
- Market Harborough website (http://www.bigfern.co.uk/)
- Market Harborough University of the Third Age (http://www.mhu3a.org.uk/)

Article Sources and Contributors

LE_postcode_area *Source*: http://en.wikipedia.org/w/index.php?title=LE_postcode_area *Contributors*: BD2412, Bearcat, Bluegamma, Dpaajones, El.Bastardo, G-Man, Georgethe23rd, Guinness2702, Kaybee2, Keith D, MRSC, MapsMan, Outrune, Palmiped, Pumpmeup, Saga City, Sajgill, Sitethief, Snigbrook, Stemonitis, TDF92, Tdoublenineone, TheFearow, WOSlinker, 19 anonymous edits

Northamptonshire *Source*: http://en.wikipedia.org/w/index.php?title=Northamptonshire *Contributors*: 1jord2, 80.255, Aaronjhill, Acalamari, Ahoerstemeier, Akadruid, Al Silonov, Alansohn, Amazonien, Ameliorate!, AmosWolfe, Andres, Andrew Dalby, Andrew Norman, Andycjp, Angr, AnnaFrance, AnnaP, Anwar saadat, AssociateAffiliate, Autoerrant, Bcasterline, Bigger digger, Birchington, Bleaney, Bogbumper, Bornintheguz, Brandon, Brookie, Brool, BrownHairedGirl, Bryan Derksen, Bwithh, CLW, Caponer, Capricorn42, Carlwev, CarolGray, Chanheigeorge, Charles Matthews, Childzy, Chris the speller, Chrisieboy, Citterio, Cj1340, Clear air turbulence, Clio64B, Cmdrjameson, Cnyborg, Cphaff, Craigsteele2001, D6, DH85868993, Dale Arnett, Dallan72, DaventryCobbler, Deflective, Delusion23, Deror avi, Deville, DinosaursLoveExistence, Discospinster, Dn9ahx, DoctorW, Docu, Dpaajones, DrKiernan, Duncan Keith, Dusimpson, Ehrenkater, Eopsid, Epbr123, Erik Kennedy, Flatterworld, Foreverprovence, Fourohfour, Francs2000, Fuhghettaboutit, G-Man, Gaius Cornelius, Gene Nygaard, Giano, Grev02, Ground Zero, Grstain, Gsingh12345, Hankwang, Happy Radio, HennessyC, Heron, Hiram K Hackenbacker, Howard Alexander, Ilikeeatingwaffles, Iloveaqua, Interplanet Janet, Iridescent, Istw, Iulianu, JBellis, Jamesinderbyshire, Jamie Mercer, Jatrius, Jeffthejiff, Jeni, Jim 21, Jimbo online, Jimly9, Jimscotland, Jjustyy, Jklin, John Maynard Friedman, John of Reading, Joshua Issac, KathrynLybarger, Keith D, Keith Edkins, Ketiltrout, Koavf, Kubigula, Kudpung, Kummi, Kwamikagami, Lampman, Lang rabbie, Laurel Lodged, Le Creillois, Likelife, LilHelpa, Lisagosselin, Lozleader, Luckystars, MJCdetroit, MRSC, Majabl, Mark J, Mark Weaver, Marknew, Mattis, Max Naylor, Mel Etitis, Miaggi, Michael Glass, MikeHobday, Modal Jig, Mohan46, Morwen, Mrbusy, Mswake, Netkinetic, Niall200, Nick Number, NigelR, Nightkey, Nilfanion, Nono64, OldSpot61, Optimist on the run, Orfeo, Owain, Oxymoron83, Packeted, Palica, Paul-L, Peterlewis, Pharillon, Philip Baird Shearer, Picapica, Plasticup, Plucas58, Prsephone1674, Psivers, R. A. C., Rauyran, ReformatMe, Rehsok, Renata, Rich Farmbrough, RichBD, Richardguk, Rkilpin, RobertG, SE7, SJB004, SP-KP, ST47, Saga City, Sam Blacketer, Sam Hocevar, Sam Korn, Sarumio, Saxontown, Secondarywaltz, Sfan00 IMG, Shove2001, Silverhelm, Sitush, Sjorford, Sqwawk, Stephenb, Swithlander, Tassedethe, Template namespace initialisation script, TenPoundHammer, Teratornis, Tesi1700, The omaga, TheGrappler, TheLetterM, TheObtuseAngleOfDoom, Thu, Tim Q. Wells, Tom-, Tpbradbury, TrulyBlue, Vegaswikian, Wellington, WelshGP3, Werdan7, Wikinator XP, Woohookitty, Woollven, YellowMonkey, 226 anonymous edits

Corby *Source*: http://en.wikipedia.org/w/index.php?title=Corby *Contributors*: Acalamari, Adam McMaster, Adam mcmaster, Adambreeze, Addshore, Ajcowan, Alan Liefting, Alexf, Alista1r999, Amertner, Amgmichael, AmosWolfe, Andrew Gray, Andrewjlockley, Andy Smith, Andycjp, Ark25, Arrokai, Ashadeofgrey, Average Earthman, Babylon77, Bad nightmare, Ben Ben, BenOWillock, Berean Hunter, Bevo74, Bhargav123, Birchington, Boxingmad, BradUK2003, Brendan1993, Brewskater, Brookie, Caltas, Canderson7, Capricorn42, Captmonkey, Cerberus 81, Chaeo, Chris G, Chris nelson, Closedmouth, Cnyborg, Cometstyles, Corb78, Corbysteve, Crazzerr, Cross2k6, CrossHouses, Crossman74, Cst17, DJNW, DMacks, DVD Smith, Dallan72, Danielle.charles, DavidLevinson, DinosaursLoveExistence, Discospinster, DjD3M, DotComCairney, Duncan Keith, Dungateshaw, EastMidsLover, Eggd2, Elemesh, Eopsid, Epbr123, Exile, Felix Folio Secundus, Fifaworld07, Fukubou, G-Man, Gerbis, Get orf me bloody bag!!, Giftedbloke, Giraffedata, Gurezaemon, Guy M, Gwd.esq, HaroldPGuy, Hmains, Hogyn Lleol, Ian Pitchford, IanOfNorwich, Iridescent, J.delanoy, JD554, JaGa, Jandgi, Javit, Jeffthejiff, Jeni, Jimscotland, John Maynard Friedman, John of Reading, Johnjoseph, JustAGal, Kate, Keith D, Keith Edkins, Ken Gallager, Keristrasza, Kievfriedcurried, KingswoodSchool, Krzysztof7260, Kusma, Kyle939, LG02, Lewisskinner, Likelife, Loganberry, Logical Fuzz, Lottiewoo, Lupin, Luwilt, Lynbarn, MBisanz, MRSC, Macaddct1984, Mais oui!, Manormadman, Marek69, MartinUK, Materialscientist, Medmedmedmed, Mentifisto, Michael Greiner, Mightysteed, Mikerobbets, Mild Bill Hiccup, Mills77, Mintguy, Morwen, Mralston, Mrevilbreakfast, Mustang Poole, Nancy, NawlinWiki, Nedrutland, Nick, Nightside eclipse, Nihiltres, Nono64, Northmetpit, Nucleusboy, Oj200, Oliver Chettle, Orioane, Outrune, OverlordQ, Palmiped, PamD, Paste, PaulBaldowski, Peanut4, Peter, Peterkingiron, Peterthecounsellor, Philip Trueman, Pinethicket, Pit-yacker, Poeloq, Pressforaction, Punkrocker1991, Quantpole, RHaworth, Radak, Rayrodden, Rcsprinter123, Rich Farmbrough, Roadrash21, Rohful, Rolo Tamasi, Russelltown, SJP, Sam Blacketer, Scarian, SchuminWeb, Scientizzle, Scope And A Beam, Scottnorthants, Search4Lancer, Severo, Shantavira, Signalhead, Silveraudi2, Sjakkalle, Sjorford, Smalljim, Smells Like Cheesy Wotsits in ere, Smjg, Snowolf, SpoonReborn, Spute, Squids and Chips, Stemonitis, Stephenb, Stickanta, Strathdon, Tagishsimon, Template namespace initialisation script, The Singing Badger, TheParanoidOne, TheST, Thombo1, Thryduulf, Tide rolls, TigerShark, Tjnewell, Tom walker, Tom-, Tonythepixel, Trident13, Vanished user 39948282, Vegaswikian, Versus22, WOSlinker, Warofdreams, Wereon, Wiki alf, WikiBomber069, Wikinator XP, Woollven, Xxanthippe, Xxxeditxxx, Ynhockey, Zantolak, Zomno, 484 anonymous edits

St._Mary_the_Virgin,_Brampton_Ash *Source*: http://en.wikipedia.org/w/index.php?title=St._Mary_the_Virgin%2C_Brampton_Ash *Contributors*: Davecrosby uk, Malcolma, Otisjimmy1, Pastor Theo, Smallerjim, Vasquez109, 7 anonymous edits

A427_road *Source*: http://en.wikipedia.org/w/index.php?title=A427_road *Contributors*: Chevymontecarlo, EastMidsLover, J04n, Jaraalbe, JeepdaySock, Jeni, Mauls, Nono64, PeterEastern, Regan123, Saga City, Severo, Snowmanradio, Tagishsimon, Vasquez109, WOSlinker, 1 anonymous edits

Desborough *Source*: http://en.wikipedia.org/w/index.php?title=Desborough *Contributors*: Acalamari, Adrian.wallis, Alanwindow, Albroun, BWDuncan, Babylon77, Berean Hunter, Bridgetfox, CarolGray, Cleduc, Cool-yeti, Cranej95, Dallan72, Darrenrht, Escape Orbit, Fantasticchris, Fifaworld07, Gfrewq12, Graymornings, Grutness, ISTB351, Ididnotthrilljfk, Jcspurrell, Joseph Solis in Australia, Jrorem, Keeno, Keith Edkins, LilHelpa, Lupin, Lynda2007, LyndaSharpe, Madmensking, Marek69, Martindesboro, Ms2ger, Nono64, OllieFury, Paidgenius, RJBrattle, Rjwilmsi, Saga City, Scottnorthants, Sintaku, Some jerk on the Internet, Steinsky, Tijuana Brass, Veledan, Warofdreams, Washouse, Wereon, Woohookitty, XKronikz, Z3ek, 63 anonymous edits

Kettering *Source*: http://en.wikipedia.org/w/index.php?title=Kettering *Contributors*: Acalamari, Adrian.wallis, Ah1954, Ahoerstemeier, AmosWolfe, Anaxial, AnnaP, Annapiranty, Anne Tate, Antisora, Ary29, AssociateAffiliate, Astatine, Asterion, Avenged Eightfold, B, BRG, Bachrach44, Bad nightmare, Bigup sim, Birchington, Bob Burkhardt, Bob Re-born, Bobo192, Brendan1993, Brendanconway, BrownHairedGirl, BulldozerD11, Bunnyhop11, Cageyh, CambridgeBayWeather, Captain Nemo III, CarbonRod85, Carl666, Cgm76, Chairboy, Charlesdrakew, CharlieA, Chris nelson, Chrisn-6, Cj1340, Colonies Chris, Crossman74, DMacks, DWaterson, Dallan72, Dancter, DanielRigal, Danielwashere92, Daveere2007, DavidJJJ, Deafrejection, Denisarona, Deville, Doberman Pharaoh, Dpaajones, Dwhdwh, Eopsid, Escape Orbit, Escapingbeaver, Fantasticchris, Fieldday-sunday, Fifaworld07, FlyingToaster, Fortnum, Fourohfour, Francs2000, Fratrep, Fuhghettaboutit, Fæ, G-Man, Geniac, George Ponderevo, Gilesmorgan, Gnevin, Goflyblind, Grahampitt, Grev02, Grutness, GuruJason1, GuruJason2, Gwernol, Haywire4, Hitchhiker89, Insanity Incarnate, Iridescent, J Milburn, Jan eissfeldt, Jeremy Bolwell, John of Reading, Jza84, KTFCdan, Karenjc, Kate Ansdell, Katie m 08, Keith D, Keith Edkins, Kievfriedcurried, Kktor, Krzysztof7260, Kubigula, Kudpung, L Kensington, Lancia123, Larry0313, Les woodland, Lewisskinner, Lightmouse, Likelife, LilHelpa, Lukasz Lukomski, LuoShengli, Lupin, M-le-mot-dit, MER-C, MRSC, MRacer, Maboughey, Majorly, Malcolmxl5, Mattgirling, Mayalld, MilborneOne, Mitchelltd, Mnemeson, MoO, Montys1954, Moonriddengirl, Morwen, Mtiedemann, Möchtegern, NVAM, Nedrutland, Niallmore09, NinetyCharacters, Nono64, Northantset1, Orioane, Pantaraxia, Peasantwarrior, Peterinns, Philip Trueman, Pinethicket, Pluma, Professoreugene, Proofreader, Prunesqualer, R'n'B, RA1678, Refsworldlee, Rich Farmbrough, Richard Harvey, Richardrox, Rohful, Roleplayer, SE7, Saga City, Sally180, Scope And A Beam, Serein (renamed because of SUL), Shabanski, Shadowpaul89, Silveraudi2, Slightsmile, Smurf101, Snigbrook, Sophie means wisdom, Sophieclarkeohyeaboi, Springnuts, SteinbDJ, Steven Walling, Stufletcher72, Tabletop, Tassedethe, TenPoundHammer, TheParanoidOne, TheSmuel, TheTrojanHought, Theroadislong, ThisIsAce, Tiptoety, Tom612pl, TomPhil, Tommy2010, Tony Sidaway, Vanished user 39948282, WOSlinker, Warofdreams, Wayneormshaw, Wereon, Woohookitty, X201, Ytny, Zigger, 451 anonymous edits

Market_Harborough *Source*: http://en.wikipedia.org/w/index.php?title=Market_Harborough *Contributors*: 1990bacon, Acalamari, Acather96, Achangeisasgoodasa, AdeleBeeby, Amgmichael, AndrewHowse, Andrewferrier, Arch dude, Arpingstone, Badgersocks, Bearcat, Berean Hunter, Bobblewik, Bobo192, Brad, Bush-14, CIreland, Cadis98, Chasedspleen, Chris the speller, ChrisGualtieri, Cj1340, Colonies Chris, Corpx, Ctopham, Cubwolf, Dallan72, Dalliance, Danmarsh, Danno uk, Daradiridatumtarides, Davecrosby uk, DeFacto, Dougielx, Earthmother1975, Edward, Epitaf, Fleckneymike, G-Man, Georgette2, Gregrouse1, Ground Zero, Hmains, Homerthecat, Intyra, Iohannes Animosus, JAMALMONK, JASADA, Jack1279, Janeiliffe, Jeni, Jeremy Bolwell, Jeremy68, John Hill njc, JustAGal, Jza84, Kaiba, Keith Edkins, Kev747, Kicior99, Kubigula, Lethaniol, Liam G-Veronica B.O.W, Likelife, LilHelpa, Lildude30201, Lofty, Lollypop0305, Lupin, MH29, MRSC, Marchant100, Marchie50, Martarius, Mclazy, MightyWarrior, Mike Walter, Mikenlunmimi, MonkeyMumford, Morwen, NawlinWiki, Neggiem01, Neilc, Nono64, Orioane, PhilKnight, Philip Trueman, Pit-yacker, Plasticspork, Plastikspork, Puchiko, Rapidconfusion, Rjwilmsi, Rohful, Ryoutou, Saga City, Samu8l, Scottnorthants, Shove2001, Signalhead, Sixty Five Pontiac, South's Captain Dave, Spoints31, StaticGull, Tenebrae, The braine, Thingg, Thinking of England, Tikki-boy, Timguk, Tommoxyz, V1459, Victuallers, Viewfinder, WOSlinker, Warofdreams, Wellington, Wereon, Xiong Chiamiov, YUL89YYZ, Yellowpurplezebra, Ylem, Yohan euan o4, Zebs, ‫היה רנ‬ ., 242 anonymous edits

Image Sources, Licenses and Contributors

Printed by Books on Demand GmbH, Norderstedt / Germany